Cosmological
Koans

Cosmological Koans

A Journey to the Heart of Physical Reality

ANTHONY AGUIRRE

W. W. NORTON & COMPANY

Independent Publishers Since 1923

New York · London

For information about permission to reproduce selections from this book, write to
Permissions, W. W. Norton & Company, Inc., 500 Fifth Avenue, New York, NY 10110

For information about special discounts for bulk purchases, please contact
W. W. Norton Special Sales at specialsales@wwnorton.com or 800-233-4830

Manufacturing by Sheridan Books, Inc.
Book design by Chris Welch Design
Production manager: Julia Druskin

ISBN 978-0-393-60921-9

W. W. Norton & Company, Inc., 500 Fifth Avenue, New York, N.Y. 10110
www.wwnorton.com

W. W. Norton & Company Ltd., 15 Carlisle Street, London W1D 3BS

1 2 3 4 5 6 7 8 9 0

CONTENTS

Part 3
Torn Apart and Reassembled

Part 4
Lofty Peaks with Endless Views

Part 5
Who Am I? Don't Know!

Part 6
Form Is Emptiness; Emptiness Is Form

A MAP OF THE JOURNEY

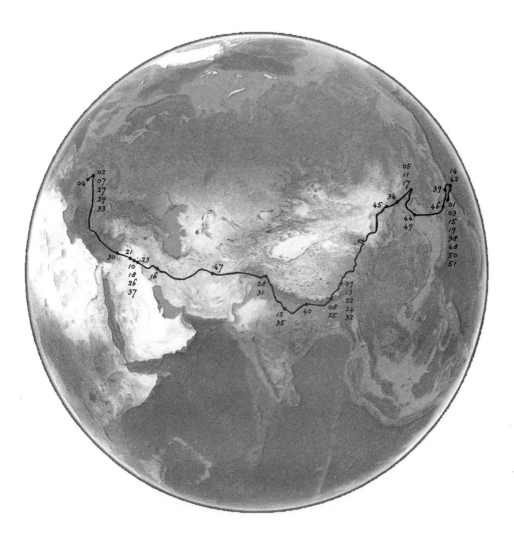

INTRODUCTION

One afternoon some years ago, I was walking through the snow and thinking about other universes. This goes a bit beyond the minimal job description of a professional cosmologist, but pondering at least *our* universe is definitely a requirement, and on this particular day I was thinking about other ones. More specifically, I was turning over in my mind the fact that the hospitality provided by our universe depends on many extremely special things. For example, if the electric repulsion between protons in the nuclei of atoms were just a bit stronger, then those atoms, and hence chemistry, and hence *life itself*, could not apparently exist. And there are many other such "coincidences." I had convinced myself that there were four—and pretty much only four—possible explanations for the fact that the laws of physics seem to be carefully chosen to allow us living, conscious beings to be here.

First, perhaps the laws of physics really were designed for us: when the universe began, it (or some superbeing that created it) had us—or at least life—in mind. Second, perhaps it was just an immense coincidence: there was one "roll of the dice" that specified, among other things, the force between protons, and we just got colossally lucky. Third, it could be that *many* "universes" exist with different laws of physics, and we are perforce in one of the universes that allow life. Fourth, perhaps the coincidences are illusory: perhaps life would somehow find a way to arise in *any* universe, with any possible set of physics.

This bit of thinking pretty much stopped me in my tracks for the following simple reason: any one of the four explanations was rather mind-boggling and exciting. Yet one of them, it seemed to me, *had* to be essentially correct. There weren't any other fundamentally different possibilities to be found. A feeling grew in the pit of my stomach that the Universe really is a pretty mysterious place.

It also struck me that this was not so much a puzzle about the Universe itself, but one that only arises with force when I—or you—ask the question. The mystery is not about why the Universe has some particular properties rather than others, but about the connection between those properties and our very existence as living, conscious, beings contemplating those properties.

The aim of *Cosmological Koans* is to explore this strange hinterland that arises between the deep structure of the physical world—from the infinitesimal to the largest cosmic scales—and our personal, subjective experience as inhabitants of that world. It is an invitation to explore deep questions in physics through personal experience, and, I hope, to gain some of the sense of mystery, excitement, and wonder that this exploration has brought me. I've devoted much of my life to physics because it represents a path by which some of humanity's finest minds have approached fundamental questions about how the world works. We are part of that world, and sometimes my own studies bring me experiences, such as that walking in the snow, in which I directly experience the mystery that underlies our world and our lives as part of it. Yet, many people—even some professional physicists—have a different experience of physics: as a rather difficult, abstract, and dry subject having little to do with actual experience, and very little to do with beauty and mystery. Even those fascinated with physics, and its exotica of black holes, time travel, quantum paradoxes, and cosmic questions, often consider these to be strange and interesting issues about the world "out there," studied and understood only by other faraway people who are, if you will, part of the high priesthood. But the beautiful unity of the world—and the physics that describes it—means that we are more intimately connected with such things than most imagine.

This book puts forward a rather radical proposal: that not only are you intimately connected with the Universe on the largest scales, but you are central. This is not to deny that you are in some sense an infinitesimal arrangement of dust on one small planet out of billions of trillions in our observable universe, which may well be one of many universes. In this physical sense you are, indeed, ridiculously insignificant. But I will try to convince you that you are also a giant: that you—a thinking, conscious being—are part of the community of such beings responsible for giving meaning, and even existence, to the universe you inhabit.

Sometime after my experience thinking about multiple universes, I was recounting it—and my thoughts about that experience—to a good friend who happens to be a longtime practitioner of Zen Buddhism. He noted that my experience reminded him of what occurs during Zen koan practice. After we discussed this for a while, it seemed clear that koan practice, while somewhat different in subject matter, was very much along the lines of the process I was looking for. Zen koans as collected in books are vignettes of a sort, that embody teachings about reality as explored by Zen adepts; there are classic collections you can buy at the bookstore. But as *practiced*, koans were devised as a means by which a teacher can confront a student with a situation that, while initially baffling, can be resolved when the practitioner is able to break through habits of thought and answer the koan in a way that is based on new insight, rather than knowledge or previous experience. And engagement with a koan is always a fully *personal* and *participatory* experience.

So, I decided to create a set of *cosmological* koans to explore the connections between us and the unimaginably immense, astoundingly complex, and ultimately mysterious Universe.

The intention of this book is not to compare, equate, or conflate physics and Eastern mysticism, as some books have done. There are real parallels that will emerge, but the adopting here is primarily of method and attitude rather than of content. In both physics and Zen, deep understanding is achieved by breaking through the previous patterns of thought and looking at an issue in a genuinely new way. These patterns can be incredibly tenacious and hidden. For example, Aristotle stated what pretty much everyone believed (and many still do): that objects tend to be at rest, move only when pushed or pulled, and return to rest when the pushing or pulling stops. It took fully 2,000 years for Galileo and others to discover (and convince the world) that the true tendency of objects is to *stay in motion*. Things come to rest not naturally, but by the action of a specific force, such as friction.

The history of physics is rife with such revolutions in insight: Newton's understanding that the physics of gravity on Earth and in the solar system is the same; Einstein's radical revolution in our understanding of space and time; Heisenberg's discovery of intrinsic uncertainty in fundamental physics, and so on. How were these leaps achieved? Much as in Zen practice, it was by being willing to sit through long hours of confusion and "not knowing," often wrestling with apparent paradoxes. And in the end, by finding the courage to perceive something in a new way. As the philosopher Arthur Schopenhauer put it: "The task is . . . not so much to see what no one has yet seen; but to think what nobody has yet thought, about that which everybody sees."[1]

By many accounts, Zen is largely about direct insight into the relation between the self—the inner, subjective world—and reality. This relation is also the subject of this book, though from a rather different standpoint of modern hard-won scientific understanding arrived at via intense intellectual effort.

Thus, my aim is to co-opt a tool and a method—the koan—to gain some real understanding of the structure of physics and what it tells us about the true nature of our physical world. This approach will also lead to questions on the frontiers of scientific and philosophical understanding that are some of the most exciting questions humanity has come to ask. But this, however worthwhile and important, is a means toward the core aim of the book: to confront your *subjective* experience of the world with the *objective* world that is described so well by physics—and revealed to be very different from what it seems. Most importantly, I hope that the book brings you, as it has brought me in writing it, experiences of the type described so well by Einstein: "The most beautiful and deepest experience a [person] can have is the sense of the mysterious. . . . To sense that behind anything that can be experienced there is something that our mind cannot grasp and whose beauty and sublimity reaches us only indirectly and as a feeble reflection."[2]

SOME CAVEATS and considerations to keep in mind:

While the format of the book is loosely based on a book of Zen koans, these are not actually Zen koans, and no experience with Zen practice or thought is necessary!

The koans make up a fictitious journey, which also has something of a story line that will come into more focus as the book progresses. They should be read as historical fiction, or parables: while real historical figures are involved at roughly the times and places they existed, most of the events depicted never took place, nor probably could have.

By contrast, I have made every effort to convey the core physics and cosmological ideas accurately, if perhaps unconventionally.

This treatment is cumulative and interlocking, and assembles various arguments and lines of thought for which the destination will not be clear until several koans or even sections later. So you'll probably want to read the koans in order, rather than jumping around.

Note that while I have minimized mathematical content, I have not eschewed equations entirely; I've included some that, for someone with a decent understanding of high school mathematics, will hopefully simplify rather than complicate the treatment.* I've also included a smattering of technical details in endnotes, for those interested.

Finally, although many koans delve into their subject quite deeply, none of them constitute a fully fleshed-out treatment; indeed the question posed by nearly any koan could easily take a book, if not a lifetime, to address. So this is not a textbook, nor is it really a full explication of any particular set of ideas. Rather (and especially in the later koans), it is a set of open doors through which you are invited to walk. This is also a suboptimal choice of books if you are primarily looking for easy answers, or indeed definitive answers at all! My view has always been that the *questions* are the important things. Answers to deep and subtle questions can be satisfying, but only briefly and incompletely, and are best viewed as useful in posing deeper, subtler, and more interesting questions. If this book succeeds, you will understand much more yet also far less of what there *is* to understand, than you do now. So if you feel confused, I invite you to see that as an opportunity and even a joy, rather than a problem—and certainly not a deficiency. I virtually guarantee that whatever you are confused about, some of humanity's finest minds wrestled for years to achieve (or fail to achieve!) some understanding of it.

---

* If you don't believe that's possible, try looking at Newton's most famous work, the *Principia*, which uses largely words rather than equations!

PART 1

The Path Laid Out before Us

What has been moved is not moving.
What has not been moved is not moving.
Apart from what has been moved and what has not
   been moved,
Movement cannot be conceived.

—Nāgārjuna, *The Fundamental Wisdom
of the Middle Way* [1]

1

THE ARROW

(KYOTO, JAPAN, 1630)

With effortless focus, Munenori Sensei smoothly pulls the arrow to bend his bow. Released like a ripe fruit, the arrow glides.

It races toward your heart.

In the eternity of the arrow's flight, you wonder: What is this present moment?

Confronting its end, your mind becomes razor sharp, cleaving time into uncountable, quickly passing moments. At one such perfect instant you see the arrow as it floats, suspended between the finest ticks of the most precise clock. In this instant of no time, the arrow has no motion, and nothing pushes or pulls it toward your heart.

How, then, does it move?

While your beginner's mind embraces the mystery, the arrow flies.

> There is such a thing as an interval into which not even a hair can be put.
>
> —Takuan Sōhō, *The Unfettered Mind*

What is this single moment, at which the arrow hangs, in midair? We often imagine time as series of strung-together instants, like the ticks of a clock. Yet if we try to focus on just one single instant of now—a single tick of the most accurate clock—we find ourselves in the company of a long line of thinkers, and start to see something rather mysterious. As William James put it: "One of the most baffling experiences occurs. Where is it, this present? It has melted in our grasp, fled ere we could touch it, gone in the instant of becoming."[2]

Let's consider an "instant of time"—this one right now, or any other. Does this instant have any duration? Does it *last* a little bit? Suppose it does last, just for a while. If so, then like any interval of time it must have a beginning, a middle, and an end. Well then, let's divide it in the middle into two shorter instants. For each of these we can ask the same question again: Does it *last*? Because we can repeat this reasoning indefinitely, it seems that we must arrive at one of two possibilities. First, that there really is no limit to how short a duration of time we can imagine, suggesting that we approach an ideal, perfect instant of time, with exactly zero duration. Or second, that we might, during our endless divisions, come to some segment of time that has some duration, yet nonetheless cannot be further decomposed: an "atom" of time.

Yet both possibilities—instants of zero duration or of finite duration—are perplexing.

Suppose the instant has, strictly, zero duration: nothing at all has any chance to happen during the instant. Then, during that instant, the arrow is only in *one particular place*. It is fixed, hanging in midair. But if it is really only in one place during that instant, then the arrow cannot possibly move during that instant, any more than a photograph of an arrow can move on its page. Motion requires going from one place to another place, but there is only one place at that instant. But now the problem is clear: if time is just many instants strung together, and during any one instant the arrow is fixed in place, then how does the arrow ever get anywhere?

This line of reasoning might convince us of the alternative possibility: that what we call an instant of time does go on for a little while, yet is discrete and indivisible—rather like the still images that are strung together to form a movie. In this view, as we imagine a movie of the arrow's flight, the position of the arrow changes from one frame to the next; only when the frames are strung together do they create motion. Yet when closely examined, this idea seems no less perplexing. Movie frames are separated by a fraction of a second, with our mind stitching them together into motion; what would stitch together atoms of time? A movie can be played at many speeds, or it can be left in its canister. If the world is like a movie being played, what is playing it, and at what speed? What prevents everything from happening at once? And what connects one frame to the next? A movie might contain an arrow in one image, and a sudden cinematic cut to the arrow's target in the next. Yet reality never does this: each inexorable moment appears to be born smoothly from the preceding one.

In short, how does motion occur, if time is composed of moments and each moment is devoid of motion? This paradox (along with several others) was put forth 2,500 years ago by Zeno of Elea, as recounted by Plato in his *Parmenides*, and by Aristotle in his *Physics*. This paradox might really bother you—if so, let it! Or, you might not quite see the problem—think some more! Or, you might feel an almost overwhelming desire to think about something else—resist it! Or, you might dismiss it as already solved, or "empty philosophy"—but such dismissal is just like walking past that small overgrown path into the woods; where might it lead? Indeed, described by Bertrand Russell as "immeasurably subtle," for two millennia Zeno's paradoxes have been tackled and explored by some exceedingly subtle thinkers, and solved—again and again and in different ways!

Yet nature solves this paradox. Things move, and the arrow flies. And in understanding motion, humanity has made vast strides from the time of Aristotle. We can predict the hour and minute of an eclipse 50 years in the future, and aim a spacecraft so precisely that it slingshots years later around Jupiter and encounters Neptune at close

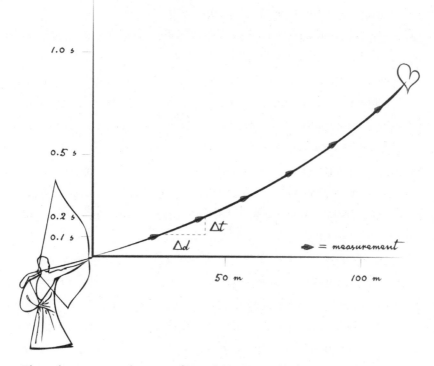

Elapsed time versus distance of Zeno's/Munenori's arrow.

range. We understand motion well enough to predict it in some situations with astonishing accuracy. How, then, does our amazingly effective physics of motion tackle Zeno's arrow?

Consider the arrow's speed. If the arrow crosses the 100 meters to your heart in a single second, we say that it moves at an average speed of 100 meters per second (m/s). Yet very careful observation would reveal that during the second half of its flight's duration, the arrow travels more slowly and covers less distance, because during its flight, friction from the air slows its motion. Perhaps it moves 55 meters in the first half second, 45 in the second half second. And we could imagine subdividing time yet further. Perhaps between 0.1 and 0.2 seconds into its flight, a duration of 0.1 second, the arrow travels 12 meters; hence, 12 meters / 0.1 second = 120 m/s during this interval.

In many ways, physics as we know it was born when Isaac Newton and Gottfried Leibniz developed the mathematics to take this idea to its logical conclusion. To illustrate their scheme for understanding motion, let's plot the arrow's *position* (in this case, its distance from Munenori) at a sequence of times, as shown in the figure at left. Very crudely, your eye measures these positions as you sense time passing in your mind, but we could imagine attaining much more precision with, for example, laser ranging and an atomic clock. In any case, through this discrete set of measurements we might draw a smooth curve that accurately describes the arrow's motion. Using this curve, we can assess how far the arrow moves over shorter and shorter durations.

This basic idea was understood even by Aristotle, but Newton and Leibniz, crucially, tackled the question of what happens when the time interval* Δt *approaches zero*—precisely the domain of Zeno's paradox. What they were able to show is that, as you would expect, during this infinitesimal time interval, the distance interval Δd vanishes just like Δt does. Yet in a well-defined and mathematically precise way, the ratio $\Delta d / \Delta t$—the speed—nonetheless takes on a particular finite and *nonzero* value. This mathematical method, which forms the roots of differential calculus, gives a response to Zeno's paradox. It says that you must not go all the way to the limit of zero duration without also going at the same time to the limit of zero distance interval—and if you do this correctly, the speed of the arrow never goes to zero. No matter how short the duration, the arrow is never at rest. There is no such thing as an instant of time during which the arrow does not move *at all*, so the premise of Zeno's paradox, as well as the necessity of thinking about "atoms of time," is neatly avoided.

Paradox solved? Perhaps. This way of looking at motion works very, very well, and we could, if we like, just leave the matter right there. But physics, like the world it describes, is a matter of great

---

* The Greek symbol Δ ("delta") often denotes the difference between two quantities. Thus, "Δt" is shorthand for "the change in the quantity t."

depth and subtlety, with secret passages and hidden rooms if you push in the right places. So let's ask some questions about the seemingly straightforward account of an arrow following its curved path to its destination.

Why this particular curve and not some other curve? The same arrow, fired again and again in the same way, under the same conditions, will follow the same path. Why? And what exactly determines *this* path out of all of the possible paths the arrow might take? (Can you imagine that the arrow actually takes all possible paths, which average out to be one that is straight, but looks curved?)

How, at some instant of time, does the arrow "know" which way to go? At that instant, it has a position, but its velocity depends on where it was a moment ago. Does the arrow "remember" where it was? Or does it carry its velocity as some intrinsic property, like its color? Why does the arrow have *inertia* that keeps it going in the same direction as its velocity, yet changes in just the right way for the arrow to perfectly trace out a preordained path?

What would happen if we tried to precisely calculate the arrow's speed at a particular instant, by measuring the traversed distance Δd during an infinitesimally short time interval Δt? Measurements in the real world are imprecise, so we can never measure Δd perfectly, nor even Δt. Then as Δt approaches zero, the speed $\Delta d/\Delta t$ will be utterly uncertain. What do we make of this? Does it even make sense to think about durations of arbitrarily short length, if we cannot know anything about motion during them? What does it even mean to measure an object's velocity? (Would you believe that just as its velocity is made up of different positions, an object's position is made up of different velocities?)

On the nicely drawn curve of the arrow's flight (see the figure on page 6), which point is "now"? Go ahead and point to one—any you like; no theory of physics will make any objection, or comment. Indeed, physics has no place, *no role whatsoever*, for this idea. Yet *you* sense "now" instantly! Try, if you like, to experience the past or future instead. You can't, can you? (Can you?) How can it be that a fundamental aspect of our personal experience has no place at all in physics?

The arrow is made up of an inconceivable number of atoms arranged and evolving together in a way that we call "an arrow in flight." What are the atoms made of? "Quarks and electrons," you might answer, or "superstrings." But whatever the finest particles are, I'll argue that modern physics suggests that they, in turn, are made of *information*. But then isn't the arrow made of information? Yes! But information about what? Known by whom or what? And how can information be pulled back, let loose, fly through the air, and pierce your heart?

These questions can be asked so quickly, in just a few hundred heartbeats. But it will be no small feat to even fully understand them, let alone answer them. So let's be on our way; the arrow is getting closer.

2

SETTING SAIL

(VENICE, 1610)

Six hundred million heartbeats earlier . . .

The fog is thick at the port, and the delay is interminable. It was not easy to leave behind so much in embarking on this insane voyage, but with the decision made, you are eager to depart. You peer around gloomily at the other boats appearing and vanishing in the fog.

As a neighboring ship slips into its berth, your heart skips a beat as you are momentarily taken by the illusion that you have set sail. You realize your mistake with amusement at your impatience. Will this journey ever begin?

Soon enough you are indeed on your way, toward the Islamic Empires and the fabled Eastern Kingdoms beyond. That night, you awaken from dreams of adventure, and feel that your quarters are all too still. Has the wind died? Are you dead in the water? Unable to sleep, you wander up to the deck and see to your surprise that the ship is quickly but smoothly slipping through the calm waters on a steady breeze.

How is it, you wonder, that you cannot tell the journey from stillness, or motion from rest?

Perhaps, you muse as you look out at the glittering stars, you and all the world are careening through space at an unimaginable speed. Would you ever know?

Shut yourself up . . . below decks on some large
ship . . . have the ship proceed with any speed
you like, so long as the motion is uniform and not
fluctuating this way and that. You will discover not
the least change.

—Galileo, *Dialogue concerning the*
Two Chief World Systems[3]

At this moment we are rotating at a speed on the order of 1,000
kilometers per hour (km/h) around the center of the Earth
as it spins on its axis. Earth moves about the Sun at about 100,000
km/h, or about *30 km/s*. In turn, the Sun orbits at 220 km/s around
the center of our Milky Way galaxy, which glides through intergalactic space more than twice as quickly, relative to the cosmic sea of
galaxies.[4] In a very real sense, we are speeding through the universe,
right now, at about Mach 1,000—the sort of speed that could whisk
someone to anywhere on Earth in less than a minute. Do you feel it?

You do not, any more than you feel the (relatively snail-paced)
1,000 km/h of smooth jet travel as you sip your drink on board, or
know, at moments of confusion in a train or car, whether you are
rolling forward or your neighbor is rolling backward. You might
have noticed this peculiarity of our world, but you probably did not
expend much careful thought on it. What happens if you do?

As Galileo did in his famous *Dialogue*, let's first delineate the
experience precisely: if we are inside a closed vessel, then our
experiences—*insofar as they are confined to the vessel and there is no
peeking outside*—are the same whether the vessel is at rest or moves
at any given *constant* speed and direction.* Confronting this experience, we face a choice.

---

* Speed and direction can be combined into a single "vector" quantity of *velocity*. As such, *constant velocity* is equivalent to "constant speed and direction." This
differs from *constant speed*, which allows for changes in direction.

First, maybe we're just not paying enough attention; perhaps there is *some way*, via a very precise experiment, that we could absolutely determine whether we are moving. For example, we might posit that human infants possess a special sense that allows them to fall asleep when a vehicle attains 100 km/h, and awaken when it again falls below this speed. But of course this is absurd; these are tiny changes relative to the 108,000 km/h at which we move relative to the Sun, so how could the infant sense one and not the other? (Presumably, it's instead the noise and vibrations that work their soporific magic.) Moreover, incredibly precise laboratory experiments spanning more than a century have all failed to find *any* effect enabling us to discern how fast we are going in an absolute way.

So, let's take the alternative route and boldly declare that there is *no possible way* to detect absolute uniform motion. If it is utterly undetectable, should we not simply abandon the idea of absolute motion altogether? But we can, of course, move, which you can prove by just doing so. So we must retain the idea of motion, but motion *relative to something else*. Put differently, two people might honestly disagree as to whether a given object is moving. Yet they would surely agree on whether two objects are moving relative to each other. A side effect of this relativity of motion is that you can always assume that *you are not moving*, even if this means that a whole lot of other stuff is moving relative to you. Every observer, in this sense, carries around a sort of "reference frame" *relative to which* they may view everything else as moving—or not moving— in various ways. This sounds a bit self-centered, but if there is no absolute reference frame, then everyone is just as entitled to privilege their own reference frame. Let's see, then, where this "relativity" leads.

As a first yet surprisingly profound question, let's ask: "If an object is in motion, is an external agent necessary to *sustain* this motion (lest the object revert to rest)?" Aristotle, as well as humanity's finest thinkers for thousands of years, believed that motion did require an external agent to be sustained. But given our bold and experientially grounded declaration that absolute motion is impossible, do *you*?

I hope not! And if not, then you have attained what was a profound insight of Galileo: if there is no such thing as absolute motion, then there is nothing particularly natural or easy or special about being at rest; it is *just as natural* to move at constant velocity! As he put it: "If placed in a state of rest, [a heavy body] will conserve that; and if placed in movement toward the west (for example) it will maintain itself in that movement."[5] Aristotle's belief that sustaining the motion of an object requires the action of an external agent was wrong.

To be fair, Aristotle had good reason to think the way he did, and the fact that humanity erroneously accepted his conclusion for so long is not surprising: it is how everyday objects act! When you shove your refrigerator, it does not effortlessly glide forever in a straight line. It stops, generally quite soon after you stop pushing it. And the harder you push, the faster it moves, so you might be forgiven for thinking that motion at constant speed requires the sustained action of a force. But it doesn't: what your push really does is to counteract the force of *friction* between the refrigerator and the floor. If the floor suddenly became perfectly slick, you can imagine the refrigerator moving far indeed, and if you can imagine a refrigerator with nothing whatsoever slowing it down, you can imagine that it might float along on its own forever, requiring nothing to sustain it. (The history of physics might have been very different, had Aristotle had more experience with large expanses of ice.)

But does the naturalness both of motion and of stillness in the refrigerator (relative to you), mean that it is easy to go between the states of motion and nonmotion? You will surely agree not! Pushing the refrigerator is hard work, and in fact, stopping a moving refrigerator would be exactly as much work as getting it going. Thus, the fridge effortlessly sustains a state of either rest or uniform motion, but resists *change* to that state. Let's name this resistance *inertia*.

It is rather astonishing that it took thousands of years for us to understand the world in this way, even though this understanding is based on analysis of the most everyday of experiences, such as the illusion that your ship is moving when it is not, or the illusion that it is stilled in the water when it isn't. What other truths are sitting there at our fingertips?

Indeed, this idea of inertia sits on the shores of a rather deep sea of thought. Let's set sail. When considering THE ARROW, we concluded that at each moment, an object has an intrinsic property called its velocity. And yet, our acceptance of the relativity of motion tells us that this intrinsic property is illusory. It corresponds to nothing: we cannot say in an absolute way whether an object has high speed or none. But while "high speed" and "no speed" are both empty of any fundamental meaning, *relative* speed is meaningful, and the *change* from one speed to another is as real as a speeding refrigerator is hard to stop! Behind this reality is inertia, which is what we call objects' ability to resist change in their velocity. Yet while familiar, inertia is strange: it measures the change in velocity, but absolute velocity is empty of meaning, and even relative velocity appears rather ephemeral, depending on which other object you happen to compare to. How do objects know how to resist change in something so ephemeral? Does an object somehow "sense" all the other objects around it, so as to know how it is moving relative to them, and resist change to this motion? How does this make sense? What *is* this thing called inertia?

Another perspective is through our *experience* of nonuniform motion. When our plane dives, our train turns, our car crashes, or our boat rocks, it is immediately obvious! When a car accelerates, one can *feel* how much, even with eyes closed, and nobody can claim that the feeling is imaginary. This is puzzling. Uniform motion is *only* relative, meaning that we can measure the velocity of one object relative to another, but cannot agree on a universal reference by which to measure everything's velocity. Yet nonuniform—that is, "accelerated"—motion can be easily and directly sensed, without apparent reference to any external object. *Relative to what, then, are we sensing it?*

What is inertia, and what is acceleration? As amazing as it sounds, nobody on Earth can authoritatively and fully answer those two questions, which inspired Einstein but are not wholly settled even by his deepest theories. And as I write this, I feel that most delightful sense of also not *really* knowing, and of adventure.

3

BEING TIME

(ZUIŌ-JI TEMPLE, JAPAN, 1630)

Sitting in this meditation hall, you've seen hours pass like days, and days pass in the blink of an eye.

Your journey to get here took you over the seas, into a cave, and through royal courts full of splendor. It brought you across forgotten deserts, down mountain passes, and through mysterious gates. Years in the tallest mountains, and in the service of a khan. Then you crossed the small sea, to learn from sages and warriors that you had not really left that terrifying cave behind, and to forge a slowly unfolding plan.

And here you sit, in stillness, artfully drifting through the years. Or do the years drift through you?

Contemplating the flow of time and reminiscing about the past, you think back to your journey's very beginning in that foggy harbor. The questions, and an old curiosity, suddenly arise: "So I'm drifting through the years. Moving in time. Steadily. But how do I feel that motion, when I could not feel the ship? If motion is meaningless, is the flow of time also empty?"

With all of your concentration on this sense of motion through time, it suddenly stops, as time floats away like a spring breeze.

As eternity begets duration, the breeze becomes the swish of the stick just before it strikes your shoulder.

Your eyes fly open to confront those of Master Zenjo. "Your heart is racing toward an arrow," he whispers with a fierce intensity. "Can it be avoided?"

> You will die! You will die! . . . Never forget about
> death.
>
> —Suzuki Shōsan, when asked by a few old women
> about the essentials of Buddhism

At some time your heart will make its last beat. A last breath will escape your lips, and you will be unable to take another. The patterns of thought coursing through your brain will fade and cease.

While you do not know *when* this time will come, and you probably dislike contemplating it, you know that you can no more avoid encountering the future than a skydiver with a failed parachute can avoid the onrushing Earth. This future is approaching you inexorably. (Or are you approaching it?) Does it, like the arrow, already exist and rush toward you fully formed, or can you dodge one fate and gain a reprieve before being taken by another?

Time is central to our human experience, yet ever so elusive to analysis. Let's think about moving through time. Do we do this? We often think this way, certainly: first, one is at some place at 1:00 pm; at the different "point in time" of 2:00 pm, one is also at some place in space, perhaps different from the first. So simple in words, yet immediately confusing: when contemplating THE ARROW, we characterized its motion as a change in position over a span of time, but then moving through time would be . . . a change in time during a span of time? What does that even mean?

Yet it seems that surely something moves: now it is 1:00 pm, and later, "now" is 2:00 pm. Does "now" move? What does *that* mean? "Now" always seems to be right here (as it were)—where would it go? Perhaps, rather, it is time that flows through now: the future approaches, becomes "now," and recedes into the past. That's often what it *feels* like. But isn't that view a bit egocentric? We don't generally think of ourselves as standing still while the street walks down us. Perhaps the ambiguity is just like the confusion of which ship is

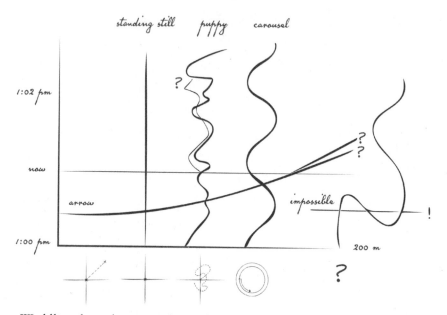

Worldline through space and time for various objects. Paths through space are indicated at the bottom.

moving: perhaps it's just as good to say that time is moving through us as it is to say that we are moving through time.

To try to create some clarity, let's return to the flight of THE ARROW. When we drew its path through space and time, it seemed pretty straightforward. At many different moments, we noted both the reading on a clock and also the position of the arrow. Each pair of readings gave us a point on our plot, and we then drew the path connecting these points. This path really represents a collection of labels: for each point on the line, there is a set of numbers—call them *coordinates*—that tell us the time and place of that point on the path. In general, we need four such numbers: three* to tell us exactly where the arrow is, and one to tell us the time. In this picture of the arrow's full trajectory, often called its *worldline*, time is very

---

* For example, longitude, latitude, and altitude on Earth. The fact that we need exactly three numbers to describe our location is precisely what we mean when we say the world is *three-dimensional*.

much like space: just one more label for "where" in *space-time* the arrow is, some way along its path.

To get a feeling for worldlines, imagine drawing such diagrams for all sorts of different objects in motion, as illustrated in the figure at left. (Because of graphical limitations, the figure displays only one spatial coordinate—say, the east–west distance from some location—along with the time.) Uniform motion would be a straight line that is more steeply sloped with slower motion, since a slower object takes more time to cover a given distance; a worldline would be vertical for something standing still. The path of the arrow is nearly horizontal because the arrow is moving very fast, and slightly curved toward the vertical at later times because it slows down. The path of a circling object would look like a wavy curve, while the curve for a puppy would wiggle all over, as puppies do.

Where is time in this diagram? We can think of it in two ways. First, there is the time *label* that we have used to know how to draw the curve, which is presumably measured by a clock ticking away in the background. But second, there is an *experience* of time that one would feel while sitting still or on a carousel, probably as a puppy, and probably not as an arrow. The distinction between these is quite familiar: as we sit reading a book, we experience time passing, but then we might also check a clock to see how much time has "really" passed. As we well know, these two notions of time can go somewhat differently, depending on the state of our consciousness. As Einstein is said to have joked: "When you sit with a nice girl for two hours you think it's only a minute, but when you sit on a hot stove for a minute you think it's two hours."[6]

From this perspective, the phrase "moving through time" seems to make a bit more sense: time is laid out, like space, and as our *internal* sense of time ticks away, we experience both different places and different *clock* times. In this way, we move through space and time, both of which exist, all laid out, quite independent of us moving through them. And this is nicely in accord with the worldlines in the figure: they are all mapped out as paths through space-time, and we follow along them.

And yet, this sort of picture leaves out some pretty important aspects of how we think of time. For one, if you look at as many curves as you like for actual objects, there is something you will never see: you will never see a curve that doubles back on itself in the *time* direction, so as to encounter a given time (coordinate) twice! This means that although it is fairly easy to avoid walking into, say, a brick wall that exists at a particular position, if we choose a particular time in the future—say, the time at which an arrow strikes—our worldline *must* encounter that time. We move through time in just one direction: forward.

Another absent aspect is that if you were to draw your own worldline and label a point on it corresponding to "now," then the part of the diagram depicting the future of "now" has quite a different meaning from the part depicting the past of "now." In particular, you don't know how to draw the future part of the diagram! Even if we can't revisit a time that we've previously passed through, we certainly can know about it. In fact, we can imagine knowing everything we require to draw our diagram completely, everywhere in space, up until the present moment. Yet just as we get to "now," we become uncertain what to draw: Will you stay here or walk away? Will the puppy run this way or that? Will the arrow miss, slowed by a passing breeze, or pierce the heart and stop its beating? One would never say that events exist and are known "to the left" but are unknown and undetermined "to the right." But we think of events to the future as undetermined, while those to the past we consider fixed and immutable.

Similarly, space does not have any sort of preferred *direction*: can you think of any processes that happen "to the right" but not "to the left"? But there are lots of things that go one direction in time. For example, it's easy to tear a page of this book. But just try to *untear* that page, which is precisely what it would mean to tear the page in the backward time direction. Nope, if you actually tore the page, then you are stuck with it torn! Time runs differently into the future and into the past. This directionality is, fittingly, often referred to as the *arrow of time* that points toward the future.

These aspects of time are intimately familiar parts of the fabric of our lives. We cannot undo the past, and we can visit it only in our memory. Nor do we know what will befall us tomorrow; we can only dream, plan, and scheme. And once something is carelessly destroyed, it generally cannot be restored. Yet these limitations accompany enormous gifts. We can create things that are genuinely new in the world: yesterday a piece of music did not exist; today it does. We can choose our destiny: what I decide today can alter the course of my life. This experience of time is deeply and profoundly focused on the *present*: it is "where" in time we can choose, act, and create. In fact, since what we know of the past is based on memory, and what we know of the future is based on prediction, what else is there but the present? As the great Zen master Eihei Dōgen put it: "Each moment is all being, is the entire world. Reflect now whether any being or world is left out of the present moment."[7]

YET THERE IS another, almost wholly opposed, view of how the world operates. Consider our flying arrow. When we come to the end of its half-drawn path that ends at "now," is there really any choice as to how the path continues? It would seem not: we know quite well which arc the arrow will take, and could quite accurately complete our diagram of its path. That is, we can *predict* which path it will take, given everything we know from the bottom half of the diagram, such as the arrow's path and speed and direction, the air's density, perhaps an oncoming gust of wind. The more we know, it would seem, the better we can predict.

If we were to perfect this ability—if we could predict the path with *perfect accuracy*—then we could go ahead and draw it with complete confidence: we could not *see* the future, but we would know what it will be. And it isn't just the arrow that is subject to these laws of physics. Dropped from an airplane without a parachute, our fate is just as determined. The laws apply to *us*, even if their application is quite complicated. We might even say that what right now we consider to be the future is already laid out before us. We'll have to wait a while to see exactly what it looks like, but it is *already there*. In this view, which we might call the *eternalist* view, time is almost precisely like space: it is all laid out at once. The future, just as much as the past, already exists. The present is a sort of illusion—just one arbitrarily chosen time out of many, with no particular significance. Nothing is created, because the future exists already. The hour of the last beating of your heart is already chosen, and you move through space-time inexorably, struggle as you might, to meet it.

Can this really be the case? We should not discard the idea just because it violates our everyday intuition about how we, and the world, work. Throughout this book we'll see many things that violate intuition and yet are true.

But in this eternalist view, when we go to decide something, when we struggle and agonize over which path we should take, what are we doing? Why does it feel that we can make the right or wrong decision? Why can we have guilt, regret, or blame? Are all these illusions? If so, what is real at all?

The challenge that Zenjo whispers is a difficult one: Is time everything, or is it nothing?

4

THE TOWER

(PISA, 1608)

We might say that your whole path starts on a hot, dusty day in Pisa as you struggle up the tower stairs with a heavy iron ball. It does not surprise you by now that your mentor, Galileo, is carrying the much less burdensome wooden ball. But the ideas to which he has opened your eyes are so fascinating that a dripping brow and damp shirt are a small price to pay.

As you reach the top, Galileo announces that you will drop both balls at the same precise moment. He asks you: "Which ball do you think will hit the ground first? Aristotle says that an iron ball of 100 pounds falling from a height of 100 cubits reaches the ground before a 1-pound ball has fallen a single cubit. And indeed, as the sweat on your brow attests, the iron ball is pulled much more strongly to the Earth."

You can only nod, still out of breath. Galileo goes on: "But Aristotle's reasoning is flawed! Think carefully. You see, the iron ball is also much more difficult to move; even rolling it on level ground takes a great deal of force."

As you ponder this, he continues: "So tell me, does the greater difficulty in moving the iron ball overcome its greater attraction to the Earth, or the opposite? Which effect will win out? Which ball will actually fall faster? I'm quite sure that Aristotle never actually tried it!"

You confess that you are uncertain. Galileo nods, and signals that you should proceed with the experiment. Even when you see it, it is a bit difficult to believe: both balls land at exactly the same time, producing two simultaneous puffs of dust (though of different sizes) on the ground far below. You turn to Galileo, who is eyeing you carefully. "How is it possible," says Galileo softly, "that two such unlike objects should fall in exactly the same way?"

I . . . who have made the test can assure you that a cannon ball weighing one or two hundred pounds, or even more, will not reach the ground by as much as a span ahead of a musket ball weighing only half a pound, provided both are dropped from a height of 200 cubits.

—Galileo, as Sygratius, in *Dialogue concerning the Two Chief World Systems*

Contemplating motion and time as we were SETTING SAIL, we decided that statements about absolute uniform motion are empty. Yet, *relative* motion and *changes* in motion appear quite real. In fact, they are the building blocks of our physical world's behavior, insofar as it can be decomposed into little bits of stuff moving around under the influence of various forces. You push, and the refrigerator moves; the fridge falls, and so do you.

We are all quite familiar with these forces. We know that it takes a larger force (i.e., "more effort") to lift or move a more massive (or "heavier") object. We also know that an object released in midair will fall. We have within us a quite powerful set of intuitions for these motions, allowing us to propel or catch a projectile with ease, avoid collisions with rapidly moving, massive, wheeled objects, and so on. These intuitions remain useful day after day because of deep and precise *regularities* behind the behavior of objects. Yet through most of its history, humanity—with only handfuls of exceptions— was content to use these regularities in a rather limited and intuitive way, without subjecting them to careful analysis.

Galileo was arguably the first to investigate these regularities systematically, and through a series of ingenious experiments such as the apocryphal test in Pisa, he showed that there are *universal*, *mathematical* rules that govern the dynamics of the everyday physical world. Amazingly, the elucidation of these basic laws, while taking millennia to get started in earnest by Galileo, was essentially completed within decades, culminating in the work by Sir Isaac Newton. This set of rules became known as *mechanics* and still forms the core of our understanding of physics. Let's look these rules, then see what exactly they say about the experiments by Galileo, which not only laid the groundwork for Newton but also profoundly inspired Einstein.

Given our understanding of position, velocity, and inertia, Newton's mechanics can be summed up very concisely: the change in the velocity of an object with time, called its *acceleration*, equals the *force* on that object divided by the *inertia* of the object:

$$(acceleration) = (force) / (inertia)$$

or, equivalently,

$$(acceleration) \times (inertia) = (force).$$

From this, it immediately follows that with no force, there is no acceleration, hence no change in velocity, so the object just moves along at a constant velocity, or stays at rest if it started out that way.

These concepts are quite precise, but in some ways different from their everyday meanings, so it is worth thinking through some mental experiments to be clear about exactly what they mean. Imagine, for example, that you have a very large wooden ball that you are pushing on level ground. As you push, the ball gradually rolls faster and faster. If you let it go, it will roll for a while with constant speed, until some other force, such as friction, slows it down.[8] Now imagine pushing an iron ball of the same size in the same way. After the same amount of pushing, the iron ball is moving much more slowly than the wooden one was; its inertia is much greater, so when the same force is applied, its acceleration is much smaller. Next, suppose that there is an exact duplicate of you who, seeing you struggling with the iron ball, comes to help. You and your twin can get the iron ball moving twice as fast after pushing for the same amount of time as before; you have doubled the force, and hence doubled the acceleration.

By mathematically defining acceleration, inertia, and force, as well as his law relating them, Newton showed that the motions of objects can be calculated precisely if one knows three things: the initial position and state of motion of each object, the inertia of each object, and the force that each object exerts on each of the others. When discussing THE ARROW, we saw how we might measure objects' positions and velocities, at least in principle: We measure where they are, relative to some fixed object, to define their positions. Then we observe how far they move over a short time interval to compute their velocities. But what about their inertias and forces?

Comparing inertias is precisely what we did by comparing the responses of the iron and wooden balls to the same force. If the wooden ball accelerates 10 times as quickly as the iron one in response to the same force, we can deduce that it has one-tenth the inertia. And if we pick a single "standard" object that we define to be one unit of inertia, then we can measure the inertia of another object in these units by comparing that object's acceleration to the acceleration of our standard object. Thus, even if we don't know exactly what inertia is, measuring it (relative to some standard anyway) is not that hard.

Finally, what about the forces? There are some forces that we all know quite well—for example, the force that a person can exert on an object by pushing on it, or the force of wind blowing on an object. Another familiar example is magnetism: Imagine you have a huge horseshoe magnet that you hold near the iron ball. Slowly but surely, the iron ball, pulled by magnetism, will roll toward you—and it will do so twice as fast if you employ two identical magnets rather than one. The wooden ball, though, is unaffected by your magnet. It seems that the magnetic force exerted on an object depends on an intrinsic property of the object—involving its composition, the amount of material that it comprises, and even its temperature—that might be termed its *magnetic charge*.

Finally we come to the force of gravity, which binds us to Earth's surface and causes things to fall from towers. While understanding of gravity's true nature would have to await Newton and then Einstein, Galileo understood two essential things about it. First, it pulls things down, toward the center of the Earth; we can call the Earth's ability to attract things toward its own center its *gravitational field*. Second, just like magnetism, gravity's strength depends on an intrinsic property of the objects that we could call their *gravitational charge*. When these quantities are multiplied, they yield the force with which an object is pulled toward the Earth—in other words, its *weight*. (The two concepts are separate, though; you would retain your gravitational charge but not your weight if you were removed from Earth.)

At this point we are fully armed to circle back and confront Galileo's challenge to determine which ball should fall faster—iron or wood. The iron ball is pulled toward the Earth with more force (because of its greater gravitational charge, and hence greater weight), but it is harder to move than the wooden ball (because of its greater inertia). Which effect wins?

Let's put our Newtonian mechanics to work. If force is equal to acceleration times inertia, and if objects feel a force that is their gravitational charge multiplied by the ambient gravitational field,

$$(\textit{force}) = (\textit{gravitational charge}) \times (\textit{gravitational field}),$$

then combining these two, we find

$$(\textit{acceleration}) \times (\textit{inertia}) = (\textit{force}) = (\textit{gravitational charge}) \times (\textit{gravitational field}),$$

or, equivalently,

$$(acceleration) = (gravitational\ field) \times (gravitational\ charge)\ /$$
$$(inertia).$$

This tells us the acceleration of any object, given the gravitational field, along with two properties intrinsic to the object: its response to a gravitational field (gravitational charge) and its resistance to acceleration (inertia).

The gravitational field is pretty much the same everywhere on Earth. But without further information on how the gravitational charges of objects compare to their inertias, it seems impossible to answer Galileo's question: objects that happen to have relatively more gravitational charge for a given inertia should accelerate faster, and those with less should accelerate slower.

Nature's response, then, is stunning. Galileo's experiment and its successors show us that insofar as we can eliminate other nongravitational forces, *all* objects accelerate at precisely the same rate in a given gravitational field. For this to be true, gravitational charge and inertia must be the same thing! In other words, the extra difficulty in moving the iron ball is *exactly, precisely, perfectly*[9] compensated by the extra force exerted on it by the Earth! This is not true of magnetism or any other force.

THIS AMAZING FACT sat essentially idle for 300 years, until Albert Einstein showed that there is a deep explanation for it, which requires a radically new way of thinking about space and time. Recall from the flight of THE ARROW that, absent any forces, objects move in a straight line, at a constant speed. Another way to say this is that in the absence of any forces, the path of an object is a straight line in *space and time*. To see this, let's plot the balls' trajectories just as we did for the arrow. As long as the ball rolls at a constant speed and direction, the path through space and time is straight. But if the ball speeds up—for example, because we have placed a magnet in front of it—it covers more and more distance through space for a given distance in time, and the path becomes curved; the wooden ball's path, unaffected by the magnetic force, remains straight (see the figure below, top left panel). We could also imagine the balls being blown by a strong wind. In this case, all three experience the same force, but the wooden ball, having the least inertia, is affected the most; the lead ball is accelerated least (top right panel). For gravity, the Earth pulls on all three balls to accelerate them. The difference, as Galileo discovered, is that all three paths curve in *exactly* the same way (bottom panel).

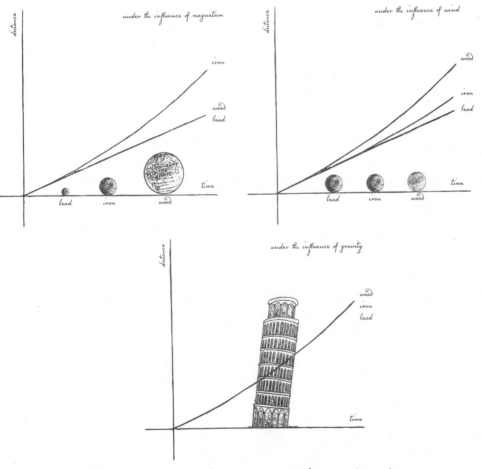

Paths through space-time with magnetism, wind, or gravity acting.

Considered in this way, forces—those things that cause acceleration in the motion of objects—really represent *reasons for objects to deviate from straight-line paths through space-time*. With no force present, the space-time path is a straight line, and the greater the force (for a given amount of inertia), the more curved the path becomes.

But we have seen that if gravity is a force, it is a strange one, because unlike any other force, it changes the paths of objects in a way that is completely independent of their mass, what they are made of, or any other property. The gravitational acceleration of an object has no connection whatsoever to what the object actually is; it depends only on the object's environment. We might dismiss this as a strange curiosity. But to Einstein it was a crucial clue to the true nature of gravity. On the basis of this clue, Einstein declared that gravity is *not a force at all*.

Wait, what? If there is no force, then why don't objects follow a straight path through space and space-time?

According to Einstein, objects under gravity's influence *do* travel in a straight space-time path!

Wait, what?

5

A PERFECT MAP

(SHENYANG, CHINA, 1617)

Quite a long way down a rather curvy path . . .

It all seemed so simple, if ambitious, at first. Among the many obsessions of Kundulun Khan is cartography, and with his boundless ambition to expand his empire, he has become maddened with the inaccuracy of available maps. Assembling his mapmakers, last month he announced: "Scholars! I wish to create a map of unsurpassable accuracy. It shall be carved into the flat stone floor of the war council building, and so precise that I and my generals may deduce with exquisite accuracy the distances between locations in my growing empire simply by measuring their distance on the map."

Following a plan of his own devising, the khan gathered a great army of horsemen, each with instruments of earthly and astronomical reckoning, and paper on which to write their observations. The khan systematically positioned his riders at fixed intervals along a line starting at the empire's westernmost edge and stretching due east. Each rider carried commands to ride north and, at each landmark, take note of the distance traversed since the last. And you? Aware of your mathematical pretensions, the khan has assigned you to aid his cartographers in the collation, comparison, and interpretation of the data.

It seemed straightforward at first, and you and the scholars were able to create excellent maps for the khan using the data. But the *perfect* map remained elusive: the more carefully you drew your figures, the more confusing and inconsistent things

became. Week after week you went before the khan to admit
failure.

Late one evening, while staring at a full moon, you have a
sudden realization: your troubles are caused by the fact that the
Earth is round, not flat! The next morning you rush to the khan,
excited by your news. But upon hearing you out, the khan, with
great scorn, exclaims: "Of course the Earth is round. If I wanted
a globe, I would have called the globe-makers here." The other
mapmakers all nodded knowingly. "I want a flat map," the khan
continued, "and thought you would be intelligent enough to
produce me one. Why should it matter that the Earth is round?
Everywhere I journey, it looks flat enough! Come back when
you have something I can use!"

You bow, chastened, and walk off. China has not treated you
well, and you miss your time in the mountains. It passed too
quickly, you think. You imagine what Tripa Dragpa would say:
something wise and reassuring like: "One step at a time. The
khan will soon understand your true measure."

Then, reflecting on your thoughts, you have it! You head
straight back to the khan.

Wait, straight?

We are all very familiar with maps and their uses, and modern
mapmaking is so advanced that we rarely have to worry
about maps' accuracy, or about how exactly they are made. But as
is often the case, this familiarity hides from us some subtle yet fascinating issues that arise when we delve carefully and deeply into
exactly what a map is, and what it does. Some of these issues are
precisely what bedeviled Kundulun Khan and his scholars, and are
also directly related to our effort to understand space and time and
movement. So, what is a map?

Most fundamentally, a map is a representation—generally
pictorial—of the land it describes, and should truthfully reflect the

relationships between elements of the territory. That is, a useful map "looks like" the territory, giving a good idea of what the territory is like and how it may be navigated. But for the khan, this was not enough; he wanted to capture the territory *mathematically* in his map, so that he could use it to measure precise distances between towns, or precise sizes of regions of the empire. To understand both his quest and his tribulations, we must think more about what makes an *accurate* map.

One way to look at mapmaking is to begin with some raw data, in the form of a compilation of the locations of every physically interesting feature of the territory. The khan's riders made just such a list when, embarking from their places along a line stretched west to east, they rode north (see the figure below). By having each rider note the distance ridden at each important feature, the khan enabled the scholars to assemble a list that gave each feature two *coordinates* specifying its location: a distance east (given by the identity of the rider) and a distance north (given by the rider's notes). These coor-

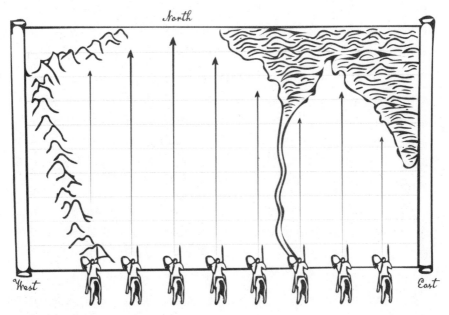

The khan's attempted map.

dinates closely correspond to the longitude and latitude used in modern maps.

But this list does not *look like* the territory. That resemblance would come by drawing each landmark on paper with a grid, where each interval on the grid represents a distance in the territory. In Kundulun Khan's example, we could draw a vertical grid line for the path of each rider, then horizontal grid lines equally spaced in distance along each rider's path (see the figure). The correspondence between real physical distances and distances on a map fixes a *scale* to the map (for example, 1 cm on paper might equal 10 km in reality). Exactly as the khan hoped, large distances in the territory could then be measured by simply measuring small distances on the paper, then multiplying by this scale.

This system is utterly familiar to anyone who has used maps, and suggests that making really accurate maps is pretty straightforward. But it isn't.[10] This would become clear if we sent a rider across the north edge of the map we just constructed. We might find that the distance between two mountains was 10 cm on the map, and thus presumably 100 km in reality. But our new rider might measure the mountains to be, in fact, 96 km apart! Something is wrong; the scale *varies* from place to place: a scale determined in one part of the map changes when you move to another part. Even worse, careful experimentation should reveal not just that the overall scale varies from place to place, but that the scale in the north–south direction is often different from the scale in the east–west direction. All this is what left the khan rather displeased, and a displeased khan is never a good thing.

But why is this complication necessary? Is there another way of making the map that would be "perfect"? No. The difficulty is in mapping the surface of the spherical Earth using a flat map; it *cannot be done perfectly*. You can intuitively convince yourself that this is the case by imagining peeling off the surface of a globe (see

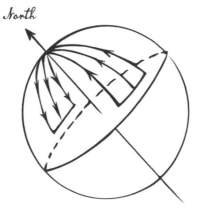

North

The khan's mapmaking on the surface of a sphere. The map will lead to a stretching of physical distances into map distances as you go north.

the figure above) and attempting to paste it onto a flat surface without stretching or wrinkling the map (representing scale variations) in any way. It can't be done, and this is exactly what yielded the distortion that afflicted the khan's map.

There are many ways of constructing world maps, and all of them comprise different decisions about which features of the globe to keep on the map. For example, you might decide that *areas* on the map should be proportional to areas on the globe. This is possible. Or you might decide that you want shapes on the map to closely resemble shapes on the globe. But you generally can't have both equal areas *and* similar shapes.

In this sense, mapmaking is telling us something crucial about the territory mapped: not just the locations of the objects, but the curvature of the *substratum—the Earth—itself.* The inability to map a surface with a single fixed scale onto a flat sheet tells us that the surface is curved. It took quite a long time for humanity to overcome its immediate intuition that the world is flat; we did so by careful investigation and reasoning. What other hidden structures and curvature might we find in our world?

———————

LET'S CONTINUE to develop our tools. Even if the territory cannot be mapped as straightforwardly as we like, with a fixed scale, can we still demand that measurements be *as accurate as we like*? Yes we can, given a bit of extra work and cleverness. The enabling insight is that if we consider a tiny region of territory—the first bit covered by a few neighboring riders, say—then our map, with its fixed scale, is extremely accurate. The variations in scale become manifest only when we compare distant parts of the map to each other.

This suggests a way to overcome the scale variation: break the territory up into little bits. The north–south and east–west scales might vary from one bit to another, but within each bit, they would be essentially fixed; and the smaller the bits, the more accurate this would be. Now imagine a curve, representing a possible path through the territory, between two locations on the map (see the figure on page 33). We'd like to know the real physical length represented by this path, but we know it is not just a multiple of the curve's length on the map, because of the variations in scale. However, let's break the path into many short segments—that is, let's take our journey one step at a time—and measure each segment's length in both the horizontal and vertical directions (see the figure on the next page). These may now be converted to *real* distances using the (nearly constant) north–south and east–west scales of the map. These two real, physical distances can be converted into the real distance represented by that short segment, which forms the triangle's hypotenuse.* Adding up the lengths of all these segments gives the total real distance between the two locations along the chosen path.[11]

This method can be used to compute distances exactly as the khan hoped—just in a rather more complicated way. (This is, in

———————

* Recall, in case it's grown foggy, that the length of the hypotenuse c is related to the other two sides, a and b, by $c^2 = a^2 + b^2$.

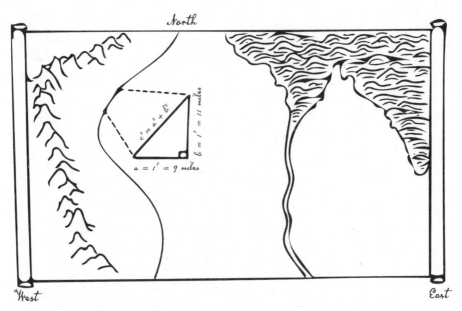

Walking the path one infinitesimal step at a time.

fact, exactly what a modern mapping program does if you ask for the length of a driving route between two locations.) We also see that the spherical geometry of the world is hidden from those who can only see nearby themselves: everyone always sees a *locally* flat world, which can have a fixed scale. But when stitching these local worlds together, necessary variations in that scale reveal the geometry of the world as a whole.

So we understand how to take our journey one step at a time to get its true measure. But if we want to walk straight back to the khan with this idea, which path do we take? What looks like a straight path on the map may *or may not* look like a straight path on the surface of a globe.

If, as we learned from THE TOWER and from Einstein, objects like to travel straight paths through space-time, then we need to know: which path is *really* straight?

6

THE COSMIC NOW

(HERE, NOW)

Right now, as you read this, a baby in India is taking its first breath, and an old woman her last. A young woman and her love are sharing their first kiss. Lightning flashes across a dark sky. The wind blows through the hair of a solitary hiker in the Sahara desert.

A satellite is seeing the Sun rise above the Earth. A hurricane is blowing endlessly through the clouds of Jupiter. Two rocks are colliding, just now, in the third ring of Saturn.

The new year is arriving on a planet around a star in our galaxy. Perhaps the world has inhabitants who are celebrating.

Our galaxy moves about 100 miles closer to our neighbor Andromeda, toward their collision and union a billion years from now.

A star in a distant galaxy ignites a titanic supernova explosion that ends its hundred-million-year life. At the same time, hundreds of new stars first ignite.

The observable universe adds enough new space for a hundred new galaxies.

All of this is happening this very second, across the universe, right now.

Yet this "right now across the universe" does not exist.

Did a child in India really just take its first breath? Probably. There are about a billion (10^9) people in India. If each person lives up to 100 years, then at least 10 million (10^7) people must be born each year just to maintain India's population, and of course the population is growing. Now, a year has 365 days of 24 hours of 60 minutes of 60 seconds—or about 30 million seconds. Thus, on average, at least one person is born every 3 seconds in India. A more careful estimate shows that one person is born about every second there, so indeed it is likely that in the last second, a baby there took its first breath.

What does this illustrate? First, that there are many interesting calculations that can be done approximately—but well enough—using just some thinking and some numbers that we might happen to know or can easily obtain. These are often called *order of magnitude* estimates. The art of these calculations is to understand the essential ingredients of a question, to know how to combine them, and to be able to obtain the result to within a factor of about 10—that is, to be able to say that about one Indian child is born each second, and not 10 per second or one per 10 seconds.

These numbers also illustrate how very large our world is. Birth—an event that happens just once in a person's lifetime—happens *every second* somewhere in the world! Similarly, the explosive death of a star as a supernova, which happens just once in the hundred-million-year lifetime of just a small fraction of stars, occurs many times per second somewhere in the observable universe.* It is a big place! (And getting bigger, to the tune of about 10^{62} cubic meters per second.)

---

* I follow the convention that "Universe" means roughly all that exists, "observable universe" means that which we can currently probe using observations with telescopes and other experiments, and "universe" means the space-time region encompassing, and with similar properties to, the observable universe. Is the Universe bigger than the universe? Bet on it!

When we contemplate all of these things happening now, it feels very intuitively clear what we mean by "now": some event either happens now, or it does not, right?

Wrong.

TO BEGIN TO SEE WHY, let's think about what we mean when we say something is "happening now," and in particular how we *know* it is happening now. When you say that a falling leaf hits the ground "now," you mean that the event coincides with your internal perception of the present. When you say that the leaf hit 10 seconds ago, you mean that your internal "clock" (or maybe a wristwatch) has ticked 10 seconds since your perception of the landing leaf.

But imagine yourself in the thunderstorm that is happening somewhere on Earth at this moment. You notice that a flash of lightning and the accompanying thunder happen at different times by your internal clock. The interpretation is simple: the thunder is a sound wave traveling at the speed of sound and takes some time to reach you; the flash, traveling at the vastly faster speed of light, takes an imperceptible amount of time. Just as for the falling leaf, this light-travel delay is so short that you generally consider the strike to happen at the same time that you see it.

On larger scales, however, the effects are quite noticeable and even dramatic. When commanding missions elsewhere in the solar system, space scientists have to contend with delays of minutes or hours between the occurrence of events and the arrival of the signals describing them. Looking at the night sky, you ponder stars as they were tens, hundreds, or even thousands of years ago. And when astronomers observe distant galaxies as they rush away from ours because of the cosmic expansion, they are looking back billions of years through cosmic history; not only were galaxies younger, but the universe itself was expanding at a different rate. We can look even farther: the so-called *cosmic microwave background* radiation consists of light that has freely traveled to us from a time 13.8 billion years ago, when it last interacted with a hot stew of hydrogen and helium gas. This light gives us an image of the structure of the universe when it was newly formed; the eventual vast array of galaxies, planets, and stars was still just an embryonic ripple in the cosmic sea. This view of our nascent universe is here with us, right now; a bit of the static on an untuned (nondigital) TV is this very radiation. In this sense we are seeing the universe's earliest epochs right now, just like the view of the falling leaf.

This commingling of the present and the most distant past makes it clear that there is a complex relationship, and there can be an enormous difference, between what is *happening* right now and what we are *observing* right now. This does not feel terribly troubling, but it is a good reason to be very cautious. In particular, it suggests that we only pay heed to things that happen at the same time *and* the same place (the same "event"). For example, perceiving a leaf "now" means that the light is reaching our eyes at (very close to) the same time and place as our internal perception of "now" is taking place; the *actual* impact of the leaf on the ground is a different event. Likewise, if you disconnected your analog TV a few years ago and had your last peek at the cosmic dawn then, this event was quite different from the initial launch of that radiation from the early cosmic fireball.

But so what? you might ask. Despite these facts about what we *know* of the universe right now, our intuition tells us that there is something like a big cosmic clock ticking away in the background, that depends neither on our perception nor on our *knowledge* of when events happened. As Newton put it: "Absolute, true and mathematical time, of itself, and from its own nature flows equably without regard to anything external."[12] That is, although we see the stars in a delayed sense, there is still *some particular way* that the stars are *right now*; we would just have to wait a while to find out. Underlying this intuition is the idea that we could have some sort of signal that travels as fast as we like, and that if we used that signal, we could see the stars as they truly shine at this very moment.

Even if no such signal exists, we can still imagine the set of events that happen right now, at least in retrospect, as it were. Suppose it is 5:00 pm, and a NASA technician sends a signal to an exploratory vehicle on Mars, instructing it to take a photo and send it back. When, at 5:20 pm, the technician receives a photo showing a waving tentacle about to grab the camera, she can infer that first contact occurred at 5:10 pm. At 5:10, though she did not know it, amazing things were happening on Mars "right now." Similarly, we could think of all sorts of events that are happening right now, by giving the corresponding definition, invented by Einstein: if we sent a light signal to an event a time *t* ago, and if that signal reaches the event, turns around, and returns to us at exactly the same time *t* in the future, then we say that the event is happening *now*. Events for which this is not true would be in the future or past. See the figure on the next page.

Taking this to its logical conclusion, by imagining signals all throughout the universe, we would think it possible to construct— in principle at least—the *cosmic now*. That is, we can imagine all sorts of events, all across the cosmos, even those that we cannot actually observe for a long time, if ever. And we feel quite strongly that each such event either is happening right now, or is not.

This construction is perfectly sensible, and accords with your intuition. Yet the universal now, the set of all events everywhere that are happening at the same moment as you read this—so clear in your mind—is just a dream.

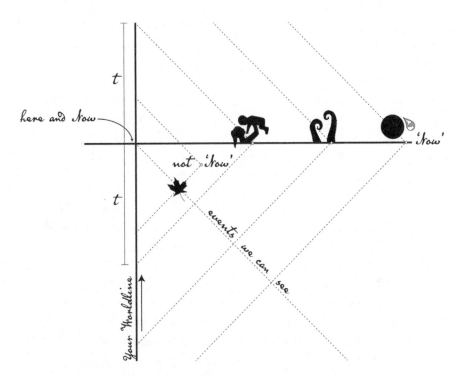

The intersection of your worldline (vertical) and "now" is here and now. Events such as the falling leaf or shining stars that we can *see* here and now are those from which light is reaching us now, and lie along the diagonal line labeled "events we can see." Events like the baby being born in India, the Martian waving its tentacles, and a planet falling into a black hole are *happening* "now," but not here; we cannot yet see them (we'll have to wait a time t, for example, to possibly see that the baby in India has been born). These "happening now" events are defined to be those events for which we can imagine a light signal sent by us a time t ago that reaches the event, turns around, and returns to us exactly a time t from now. In this way we can extend our "now" cosmically.

7

DRIFTING DREAMS OF VENICE

(VENICE, 1609)

It is a warm and lazy spring afternoon, and the swaying of the gondola as it drifts through the canals of Venice has nearly lulled you into a nap. But Galileo's indefatigable mind is seldom idle, and on sudden inspiration he produces from his pocket a small ball. "Observe!" he pronounces a bit too loudly. "As I drop the ball, it falls straight down to the floor of the boat, just as it would on land; the motion of the gondola has no effect on its behavior." This seems quite obvious, and Galileo has pointed it out before. You try to muster some alertness as he goes on: "But look at the girl there on the bridge. As I drop the ball, what does *she* see?" Your first thought is that she sees just what you do. But knowing Galileo's trickery, you take a moment, and it comes to you. "Ah," you respond, "she sees it fall down, but also forward, because of the gondola's motion."

Galileo is delighted: "Very good! Indeed the girl has a different interpretation of the same occurrence, in which the ball moves with a different velocity than we attribute to it here in the boat. In fact, there is a mathematical connection between what different observers see. For example, if I throw the ball to the girl as we are about to pass under her bridge, she sees a more robust effort than my old arms can manage; for, in addition to the velocity I impart upon the ball, she sees also the velocity of our gondola. If, likewise, I throw the ball back to her after we have passed, she sees my throw enfeebled by precisely the same amount." You consider this observation for some time, concluding that it is undoubtedly correct and in accord with all of your experience.

All that day you feel pleased at having come to understand another facet of the world. But that night you have a strange and terrible dream: You are on the river at night in a gondola, but the river is filled with other boats, with a Galileo in each one. The Galileos are equipped with glowing balls of light, which they throw back and forth to each other. But in cruel mockery of your teacher's demonstration earlier that day, no matter how hard one Galileo throws a ball to a second Galileo, and no matter what the relative speed of their gondolas, the ball arrives at exactly the same speed. You then notice an even more fantastical distortion of events: Two Galileos are playing catch with a glowing ball, but the faster their gondola glides, the more slowly they seem to move, as if caught in amber. Events swirl about in so confusing a manner that soon you can no longer tell distance from duration, or action from existence.

Waking in a cold sweat, you get up, light a candle, and reassure yourself that you are back in the normal, sensible world. You do not even consider the possibility that everything you dreamt was true.

> The eyes of samadhi can turn fire around. When the clouds fly the moon travels; when a boat goes the shore moves. It is just like this.
>
> —Shakyamuni Buddha[13]

The idea that there is *no meaning to absolute uniform motion* set us, along with Galileo and Newton, on the road to understanding the mathematical rules lying at the foundation of the familiar and everyday world we inhabit. Taking this same path further reveals ways in which that everyday world is much, much different from how it seems.

The physics that Galileo developed, which formed the foundation for Newton's, rested on two pillars. The first was exactly this equivalence of different states of uniform motion, often called the *equivalence of inertial frames*. Here, "equivalence" means that you cannot tell one frame from another using experiments you do within that frame, "inertial" means unaccelerated, and "frame" means a sort of system of reference—for example, your current surroundings, or the interior of a boat—with respect to which you can measure positions and velocities of objects. The second pillar of Galileo's physics is a very sensible rule governing how velocities measured in different inertial frames should be compared and combined. As Galileo demonstrated, if you throw a ball forward at speed v_{ball} from the bow of a gondola, while the gondola is moving at speed v_{gond} toward the bridge, someone on the bridge will see the ball move at speed $v_{ball} + v_{gond}$. This is the mathematical rule that Galileo refers to; it relates the speed of the ball v_{ball} in the *gondola inertial frame* to its speed in the *bridge inertial frame*, $v_{bridge} = v_{ball} + v_{gond}$, given the relative speed v_{gond} between the two inertial frames.[14]

When you think about this for a while, indeed it seems so clear as to be quite hard to doubt, doesn't it? But in the late nineteenth century, physicists encountered a conflict with this idea. One such physicist, James Clerk Maxwell, was studying the electric and magnetic forces as represented by *fields* that permeate space, and he had derived a beautiful system of equations that unified and governed the behavior of these two fields as one *electromagnetic* field. Maxwell's equations, most interestingly, showed that just as there can be (sound) waves in air, or waves in water, there can be *electromagnetic waves* in the electromagnetic field. And when he calculated the speed of these waves, he found that they travel at a fixed, particular speed c of about 300,000 km/s—and never at any other speed. This was satisfying, good, and troubling.

It was *satisfying* because this is just the speed of light, and Maxwell correctly hypothesized that his electromagnetic waves *were* light, thus providing both a deep explanation of light, and a connection between it and other physics. It was also *good*, because it was completely consistent with Galileo's equivalence of inertial frames: according to Maxwell's equations, whatever inertial frame you are in, if you do experiments you will find that light travels at c = 300,000 km/s, so this speed cannot be used to distinguish one frame from another. Yet it was *troubling* for just the same reason, because it *also* holds for signals sent in one inertial frame and received in another: a signal sent at speed c from within one frame will be received at a speed c in another frame, *regardless of how fast the second frame is moving with respect to the first*. This behavior is *not* compatible with Galileo's postulated transformation rule between frames. According to Galileo's rule, observers in different frames should accord different velocities to the same beam of light, just as for the ball on the gondola!

Einstein tackled this problem. The obvious solution would be to say that Maxwell's equations are not being applied quite correctly, and that light should behave just like any other object or wave. But Einstein isn't famous for doing the obvious thing. Instead, Einstein decided to reject Galileo's velocity transformation rules. He proposed, in his *principle of special relativity*, that *all* physical laws, including Maxwell's equations governing the speed of light, are the same for every inertial reference frame. That is, uniform motion cannot be detected by *any means*, including using the behavior of light. The logical consequences of this principle are strange, wonderful, and best of all, true.

TO SEE WHAT some of those consequences are, let's return to the gondola dream. All of the Galileos are equipped with the glowing balls that, like light, have the strange property of *always traveling at the same speed*—let's say 2 m/s. They cannot be slowed down, or sped up, but their direction can be changed.

Now imagine being on a gondola that is floating down the river at 1 m/s. In the gondola are two Galileos, standing on opposite sides of the boat and throwing a glowing ball back and forth in a game of catch (top panel in the figure below). Since the ball moves at 2 m/s,

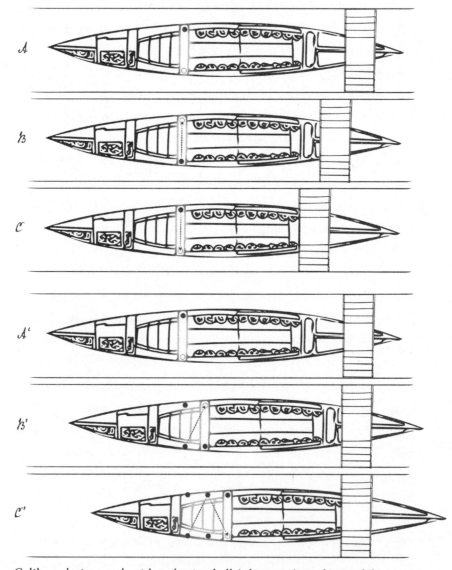

Galileos playing catch with a glowing ball (white spot), in the gondola frame (top: A, B, C) and bridge frame (bottom: A′, B′, C′). From A (ball at bottom) to B (ball at top) to C (back to bottom) forms a sequence of three events that constitutes one tick of a sort of clock.

and the two Galileos are 2 meters apart, you see the ball pass from one Galileo to the other Galileo and back again in 2 seconds, so its motion forms a crude clock that "ticks" once per second.

Now it gets tricky. Imagine disembarking, walking onto a bridge, and observing the same gondola as it passes underneath. What do you expect to see? Picture the path the ball has taken, as viewed from above. Because of the gondola's motion, this path looks like a zigzag pattern rather than a simple straight line (bottom panel of the figure). How long is each of the segments? The gondola is 2 meters wide, and there is 1 second of time between catch of the ball, during which time the gondola moves forward 1 meter. So, by the same geometry employed in the PERFECT MAP, the length of each segment squared should be the sum of the squares of the two sides, or $2^2 + 1^2 = 5$. This gives $\sqrt{5} = 2.2$ meters for each segment—hence 4.4 meters for the round-trip. Now we have a problem: We have deduced that while sitting on the bridge, we should see the ball move 4.4 meters in 2 seconds, or 2.2 m/s, but that is impossible. By our postulate, *the ball only ever moves 2 m/s.* Where could the reasoning have gone wrong?

The problem is not with the Pythagorean theorem or the lengths that go into it; if we liked, we could directly measure the distance between the positions at which the ball was caught. There is only one place left to look, which took Einstein to first realize and embrace. The weakness he found in the argument is the tacit assumption that *1 second of time on the boat is the same as 1 second on the bridge.* What if it isn't?

On the boat, the ball takes 2 seconds of "boat time" to go from one Galileo to the other and back. Yet in the view from the bridge, the ball covers 4.4 meters, moving (as it must) at a speed of 2 m/s. Thus it takes $4.4/2 = 2.2$ seconds of "bridge time" to cross the gondola twice. But since the *same thing* is happening, the only possible conclusion is that boat time is different from bridge time—that is, that anything that seems to take 1 second on the boat takes 1.1 seconds on the bridge.

The price of accepting Einstein's principle of special relativity, therefore, is that we must give up the notion of a universal time: if we consider two events, such as the throwing of a ball by a Galileo or the catching of that ball after it has been thrown back, then different observers disagree about how much time has passed between the events. This is what Einstein concluded, and he derived an exact formula to describe how quickly time passes in one inertial frame relative to the other. Put more personally and applying Einstein's formula using actual quantities, if your friend goes for a 10-minute walk (by her watch) while you stay in one place, then because she is moving while you are staying still, 10 minutes *plus 3 femtoseconds* will have passed for you when she returns.

The effect is incredibly, undetectably tiny in real life as opposed to the gondola dream, because light travels at 300,000 kilometers (not 2 meters) per second. You could live thousands of years and never, ever notice it in your everyday life. But there it lurks: a fundamental difference between the way you think the world works, and the way it actually does. And make no mistake, it is real and well tested: for example, each satellite in the global positioning system (GPS) carries an extremely precise atomic clock. The rate of time passage on these satellites is verifiably different from the rate on Earth. In fact, this and other relativistic effects must be treated carefully in order for GPS to function at all; failure to account for them would lead to errors of tens of kilometers per day in your location![15]

THESE MIGHT SEEM like tiny "corrections" that we must take into account for some technologies but are otherwise fairly irrelevant, and dismissible as a curiosity. But they have profound wider implications, as we can see if we return to the question of THE COSMIC NOW. Recall that we carefully constructed our notion of "now" in faraway places by imagining that we sent a signal a time *t* ago to some such place, and that *if* we get the return signal from that place a time *t* from now, then we define the place and time of the signal's bounce as an event that happens *now*. But let's try this in our gondola dream. Imagine a gondola with a Galileo in its center, who simultaneously throws one ball to a Galileo in the bow of the gondola and another ball to a Galileo in the stern, each 3 meters away. In the gondola frame, each ball traverses the 3 meters in 1.5 seconds, since the balls travel at 2 m/s. If they are instantly thrown back, then the central Galileo catches them both at the same time—and thus in the same event. Thus, by our definition, the central Galileo will say that the other two Galileos catch the balls at the same time— simultaneous with a time 1.5 seconds after the central Galileo threw the balls.

This all seems straightforward enough: throw two balls at an equal speed for an equal distance in two directions that bounce back, and they will arrive back at the same time. But now let's look in the bridge frame, which is equally valid. In this frame, during the balls' flight the boat moves in the direction of its bow. Therefore the ball thrown toward the front of the boat must cover a greater distance, which takes longer. Thus, *one ball reaches the bow after the other ball reaches the stern*. The two events, which are simultaneous according to the Galileo in the boat, are not simultaneous according to an observer on the bridge.

Because the bridge inertial frame and the gondola inertial frame are equally valid, we cannot say that one notion of simultaneity is more valid than the other. We don't notice this ambiguity in day-to-day life, because we move slowly relative to light, and only deal with distances over which light travels imperceptibly quickly. But the issue becomes quite significant if we attempt to extend our reference frame so as to talk about "now" over vast distances.

In the center of our galaxy, a gigantic black hole holds court. Imagine that this black hole is cruelly ripping apart an inhabited planet *right now*. Since this is far away, for "right now" to mean anything we must mentally construct an inertial frame including both us (at rest) and the doomed planet. We can then state that in this frame, the planet's destruction is simultaneous with our experience of now. But consider the exact same construction according to someone leisurely walking past us and toward the galactic center. In that frame, the planet was already gobbled down a full hour ago! *Who is correct?* Do the countless thinking beings of that civilization still exist, living their last minutes, or are they tragically and irrevocably gone?

Einstein tells you that despite your intuition screaming that it is either one or the other, there simply is no meaningful answer to the question.

The universal now is a fiction.

8

CHOOSE YOUR PATH

(HIMALAYAS, 1612)

The view from the top of the pass is breathtaking, and you pause to enjoy an endless sweep of mountains and inviting valleys arrayed under a vast sky.

Or at least it would be breathtaking if you had any breath left, and you immediately reflect that it would be much easier to enjoy if the horse hadn't just run off, or if the cart containing all your worldly possessions (and from which you just so unwisely unhitched your horse) could pull itself, or even if the road you had to drag it along were dry rather than still mucky from a freak cloudburst.

Down the mountainside lies a network of trails left by countless caravans descending the pass. Too tired to think much about which would be the very best way, you start down in a somewhat random direction. But as you work your way down, you realize this is a mistake and arrive at two important considerations.

First, the cart is too heavy for you to pull uphill any real distance. In fact, if the grade becomes too shallow, the cart tends to get stuck and be very hard to move. So there is a minimal steepness to any practical path you can follow.

Second, steep segments of the path are much easier and enjoyable to travel. But if you pick only steep segments, you inevitably run into a segment that becomes too shallow or even goes uphill. So you have to balance the easy, steep segments with more shallow ones that cover more ground.

You can see, in the distance, your eventual goal where the

paths converge near an outcropping by the river. But what path
will get you there with least effort?

Your legs are growing very tired. You realize your food was
in the horse's saddlebags, and you haven't eaten for hours. Your
back and arms are aching from the effort.

The network of complicated crisscrossing trails stretches for
miles and miles ahead down the slope. How do you choose your
path?

Choose the path of happiness!

> A poet might say that water runs down hill
> because it is attracted to the sea, but a physicist
> or an ordinary mortal would say that it moves as
> it does, at each point, because of the nature of the
> ground at that point, without regard to what lies
> ahead of it.
>
> —Bertrand Russell, *The ABC of Relativity*

The problem of getting down the mountain with the least effort
is a very common type of problem: how to choose a path
through space for which something is minimal. For example, you'll
often seek a path of minimal distance, so as to get to your destination in the quickest way possible. This effort might involve laying
out in your mind—or on paper—a number of possible paths, measuring their lengths, and seeing which is shortest. But you may soon
discover that *shortest* and *quickest* are not at all the same: taking a
fast road rather than an overgrown path may be longer in distance
but much shorter in time. To find the quickest path, you could take
each prospective path, break it into little segments of length Δd, and
in each segment estimate the speed v at which you could traverse the
segment. A given segment would then take a time $\Delta t = \Delta d/v$, and
adding up these segments would yield the time necessary to traverse

the path. Comparing this time across all possible paths would then tell you which is quickest.

The challenge in finding a path down the mountain is a bit different. Your goal in descending the mountain is to use the least *effort* rather than the least time. The constraint is that you have to cover a certain horizontal distance (to reach the outcropping by the river) as you descend a certain altitude. You can choose shallower path segments that are hard but cover a lot of ground, or steep segments that are easier but may drop altitude too fast. One way we could express things is

(effort) = (amount of descent) × (difficulty).

Here, *difficulty* represents how hard it is to descend a certain amount of altitude, and we know that the difficulty is greater for flatter path segments.

Just as in adding up the time between two points in space, we could add up the effort between your initial and final altitudes by dividing the descent into small decrements. For each decrement, we multiply the amount of descent by its difficulty, obtaining an amount of effort for that segment; then we add up all of the resulting bits of effort to get an overall amount of effort for that path down the mountain.

The pattern is clear: for both travel time and effort of descent, we divide up the interval we want to traverse, then multiply each interval by a quantity—let's call it L—and sum the product. The quantity L can depend on various aspects of the path, such as your velocity, or how hard you have to work to go down a slope of a given steepness. Finding the path amounts to adding up L over each path to get a total quantity—let's call it S—and finding the path for which S is minimal, be it travel time or effort.

It looks like the problem at hand, then, is to consider all possible paths down the mountain, break each into segments, figure out the steepness of each segment, determine how difficult that steepness would be to navigate (keeping in mind that the steepness cannot go below a critical level beyond which the cart gets stuck), multiply that difficulty by the amount of altitude descended, add up all the segments, do this for all the paths, and finally see which path has the smallest total effort.

That sounds . . . really hard. If actually faced with this situation, what would you do? No such thing! Rather, you would probably pick a path that looks like a reasonable amount of descent for a given amount of distance—a happy path—perhaps by looking a little bit ahead down the pass so as not to get stuck anywhere, and muddle along as best you could while trying to enjoy the scenery.

Wouldn't it be surprising to find, after making your way down the path in this way, that you had, in fact, chosen *exactly* the easiest of all possible paths, never having taken a wrong turn?

And yet, this is exactly what physical particles do! About 150 years after Galileo's time, Joseph-Louis Lagrange formulated a beautiful set of equations that connect the *Lagrangian* (our *L*, not coincidentally) to the *force* felt by a particle traveling between two space-time events. These equations allow us to see the same physics—the laws governing how objects move through space and time—in two very different yet equivalent ways.

In one, which we've seen already, an object at each moment experiences a force that impels it to change its velocity in a particular way; the effect of these forces over time leads to a path for the particle.

In the second, the path as a whole is "selected" by nature to be the one out of all possible paths that either minimizes or maximizes (*extremizes*) the total accumulated amount of *L*. This method, which has additional applications, often goes by the name of the *principle of least action* (though that's slightly misleading because, as we'll see, sometimes it's the principle of *most* action). This method and the method of forces and velocities yield precisely the same results. Beautiful!

But what about a particle that feels *no* forces? We've seen with Galileo that it should travel a straight path, meaning that its velocity should not change. But we can also identify the straight path as the one of *minimal spatial distance*, where we compute the spatial distance by breaking up the path into little bits and adding up their physical lengths. In this sense, if there are no forces, the Lagrangian L that is added up is simply the physical distance.

THERE'S ONE MORE step we can take. Talking about space alone is fine for Galileo, but if we want to follow Einstein, we must consider space and time together. Recall that in our discussion of THE TOWER, we also saw that an object with no forces on it travels a straight line in *space-time*. Can we identify this path also by minimizing or maximizing some quantity? Yes, but we have to be careful: the quantity must be something physically meaningful and unambiguous, *not* something like the spatial distance or the temporal distance traveled, because in DRIFTING DREAMS OF VENICE we saw that both of these are *relative*: they depend on the chosen reference frame.

Suppose that, attached to our object as it travels through space-time, is a wristwatch, or perhaps a beating heart, if the object is a person. As the object follows some path between two space-time events, the wristwatch must register a number of elapsed seconds; call it ΔT. This number is a simple fact, which nobody can disagree with, even if they have different reference frames and different senses of simultaneity and distance in those reference frames. Yet there is a *relation* between ΔT and those frame-dependent distances and durations. Einstein (and Hermann Minkowski) showed that for a small path segment, the wristwatch time ΔT could be computed in terms of the spatial distance Δd and the temporal distance Δt, in much the same way that, when forming A PERFECT MAP, we could compute the spatial distance Δd from the east–west and north–south distances. There are just two differences, albeit crucial ones. First, we must convert Δd into a time interval, by dividing by the speed of light c. Second, we must *subtract*, rather than add, this duration from the time interval Δt. The net effect is

$$(\Delta T)^2 = (\Delta t)^2 - (\Delta d/c)^2.$$

What is wonderful about this quantity is that it is designed so that it comes out the same *no matter which frame Δt and Δd are computed in*. (In fact, we can look at it the other way: this equation, in a sense, *defines* the relation between inertial frames in Einstein's special relativity theory.) This quantity can then be used just like the Lagrangian L or the spatial distance d: divide up the path a particle might take; compute ΔT along each little bit of path; then add them all up to get T for the whole path. Let's look at this approach applied to three different space-time paths (see the figure below).

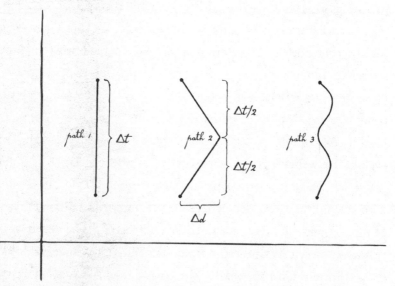

Three paths through space-time with the same starting and ending time.

The first path, on the left, is straight. The second is composed of two straight segments, with a change of velocity midway through. Both have the same total temporal extent Δt, but the straight path has no traversed spatial distance Δd to subtract. Thus, the wristwatch time ΔT for the first is greater than for the second. Likewise, the third path, which wiggles back and forth, also has a smaller ΔT than the first one. The implication is clear: any change in direction of the path leads to less wristwatch time, so *the straight path is the path of maximal wristwatch time.*

To summarize all this, the path of a particle through space-time can be determined by maximizing the total S of a quantity L that is added up along the path. This quantity L includes both the wristwatch time ΔT and other ingredients related to forces.

This is pleasing, but it is also somewhat puzzling, when you take it seriously. It says a particle can use the forces it feels, and the velocity it has, to move *right now* with utter assurance that over the course of the next year, it will trace out a path that turns out, in retrospect, to be the path of maximal action S out of all possible paths it could have taken during that year. Along that path, it will interact with many things in many complicated ways. But in the end, it will get things exactly right. How is this possible?

You are made of particles. You, and all your component particles, are choosing directions right now. Which path will you look back and realize you always have been taking?

9

TAKING THE LEAP

(GANDEN MONASTERY, TIBET, 1612)

You are peaceably sitting and reading a book, while occasionally admiring the sunset over the mountains, when Tripa Dragpa steps into your light. You look up, and the Sun forms a halo of crazy hair about his head.

"Throw that book through the air," he demands. "Throw it! Does it travel in a straight line? If you say yes, you will be on the right path but must put out your eyes. If you say no, you must jump off a cliff."

So, does it?

> I was sitting in a chair in the patent office at Bern when all of the sudden a thought occurred to me: "if a person falls freely he will not feel his own weight." I was startled.
>
> —Einstein[16]

When an object is thrown, its path appears to curve (you would have to be blind not to see that), because gravity appears to pull it down to Earth. Einstein described the thought he had in his chair at the patent office as "the happiest thought of his life" because it eventually led him to realize that what *looks* like a curved path should be considered straight.

Imagine jumping off a cliff with this book. Because you are imagining, you don't have to worry about what happens when you hit the bottom; you may also neglect the resistance of the air you are falling through—the only thing acting is gravity. You can just relax and enjoy being in free fall, able to perform an experiment. Throw the book. In light of Galileo's experiments at Pisa, will the path look curved or straight?

Let's investigate. According to Galileo's (now precisely confirmed) findings, you, the book, and anything else are all accelerated toward the Earth at exactly the same rate. When you tossed the book through the air (or imagined doing so), its path appeared to curve because from one instant to the next, the pull of the Earth's gravity increased the *downward* part of the book's velocity, whereas in discussing Galileo's experiments, we saw that the straight-line path through space was characterized by *constant* velocity.

If, however, you were to throw the book while falling, things would look different to you. In this case, just as before, the book's downward velocity continuously increases because of gravity, but *so does yours*. Thus, the book's velocity *relative to you* remains constant, and the book appears to travel in a straight line, just as it would if there were no gravity at all. This may be more clear if you consider just you and the book enclosed in an imaginary box that is falling with you: it just looks like the book is drifting away at a steady speed* (see the figure on page 62).

According to an observer watching from atop the cliff, the book appears to have a curved path, as do you. But the book travels in a straight line relative to you, who have jumped off with it. So who is right? *Is* it a straight line or not? This is quite reminiscent of the question of distance, or duration, between different events: each observer

---

* This effect can be used to train astronauts by taking a high-altitude airplane into a steep dive: if the airplane is effectively in free fall, then its passengers are weightless.

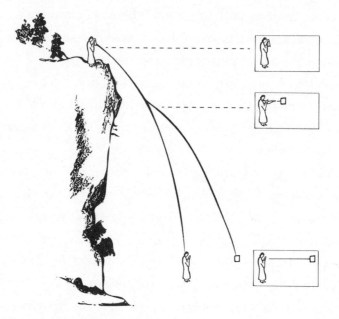

A thrown book in two different frames.

has a unique description, connected with that person's own reference frame, and all are equally valid. Yet there is also something *objective* characterizing the events' separation, which is the *space-time interval*. Can we do something similar here with space-time, and somehow reconcile disagreement even as to the straightness of a path? And if we take the space-time view, is the path straight?

Einstein said yes, and yes again, and contained in these yeses is a whole new way of looking at gravity, space, and time. To see how, we must assemble three strands of thought that we have followed thus far, along with several clues that we have picked up along the way. Let's revisit these three threads, then try to weave them together.

The first thread took us from Munenori's arrow to experiments with fridges and rolling balls, then wandering all over a well-mapped territory, and down a mountain pass. Through it we've seen that objects have a natural propensity in their motion: either to sit still, or to travel in a straight line at constant speed. This behavior can be described very simply as objects following a straight-line path through *space-time*. These paths have a particular property: just as a straight line in space is the shortest distance between two positions, the path that objects follow through space-time is the *longest* path between two *events*, if we measure the space-time length between the events by the "heartbeat time" experienced by someone following that path between the two events. Finally, *forces* can then be considered as any effects that cause objects to deviate from this natural path through space-time.

In summary, *in the absence of external forces, an object traverses a path between two events that is a "straight line," defined as the path maximizing the total heartbeat time between those two events.*

We followed the second thread by ship and gondola. Through it we learned that the "natural" straight-line motion just discussed happens in only certain reference frames. A reference frame is a sort of large-scale framework for measuring positions, velocities, and times of events, such as the room or region you are currently in, or the interior of a boat or a gondola or an airplane. We call the particular frames in which straight space-time paths are followed *inertial frames*. Given one inertial frame (say, the bridge), another frame that moves at constant speed and direction with respect to the first (say, a gondola) constitutes another inertial frame, and there is no way to distinguish one inertial frame as more special than, or preferable to, any other.

There is a particular rule for translating the description of what happens in one inertial frame into the description in another; for example, objects at rest in one frame will be moving when described in another inertial frame, in a way that Galileo described and worked out. But as Einstein discovered, the fact that *light* travels at the same speed in any inertial frame requires Galileo's rule to be modified into a different rule. This modified rule distorts space and time intervals, so that two events separated by some amount of time in one inertial frame are separated by a different amount of time in another frame. But the transformation does *not* alter the heartbeat time (i.e., the space-time distance) between two events, so all frames agree on how much heartbeat time passes between them. What about noninertial frames, such as those that rotate or accelerate with respect to an inertial frame? They are not evil, but if we describe events within one of those, then objects no longer tend to traverse a straight-line path of constant speed. Rather, they will appear to be subject to various "fictitious"[17] forces pushing them this way and that, such as the forces you feel toward the front of a rapidly braking car that a seat belt protects you from.

To summarize, *inertial frames, in which objects follow straight space-time paths, are indistinguishable; non-inertial frames are distinguished from inertial frames and from each other by the "fictitious" forces that arise in these frames.*

The third thread drew us up the tower and off a cliff, where we saw that all objects fall at the same rate, such that a (small and non-rotating) freely falling frame—even though it would seem to be accelerated—is *indistinguishable* from an inertial frame in which gravity is absent. So, gravity and acceleration are in some way interchangeable. We can expand on this idea by giving the tower an elevator and going for a ride on it. If we climb in and start to ascend, we notice two effects on us. First, gravity pulls us—and everything else in the elevator—down toward the elevator floor, just as always. But second, the elevator floor moves up toward us and everything else in the elevator. Because both gravity and the elevator's acceleration upward affect everything in the elevator in exactly the same way, there is no way to tell the two effects apart. That is, when an elevator accelerates up, it is *just the same* as if gravity were suddenly a bit stronger; when it accelerates down, you correspondingly feel a bit lighter.

To summarize, *a freely falling frame is equivalent to an inertial frame without gravity; likewise, the constant acceleration of a frame is equivalent to a constant contribution to the ambient gravitational field.*

LET'S PUT these strands together. Our major clue about how to do this is that Einstein would consider the path of the book flying across the room to be a "straight line," even though it looks curved. Indeed, the path *is* curved; if you measure various distances between the throwing and landing point of the book, the path it takes will not be the shortest. But this is according to the (by this time) rather small-minded view of thinking just in terms of paths through *space*. Let's be more Einsteinian, and think of the paths through *space-time*; for this, we must fall back on our definition of "straight" as maximal heartbeat time. At first, this methodology would not seem to help: when we considered it before, it yielded paths of constant speed and direction, which are *not* the sort of path that a thrown book follows. Yet that previous consideration had a big, and hidden, assumption. This assumption was just as important—and just as wrong—as the assumption that the khan's territory could be perfectly mapped into a flat surface.

Just as the Earth is curved, *space-time itself must be curved.*

Einstein's genius was that he saw this possibility, but he had no idea how to describe such curved space-time. Fortunately, others did. In the early nineteenth century, János Bolyai, Nikolai Lobachevsky, and Carl Gauss independently developed an understanding of how geometry works in a curved space—one in which initially parallel lines can converge or diverge. Using this mathematics, one can create a map (otherwise known as a *coordinate system*) describing the surface, as well as a sort of scale (technically known as a *metric*) that allows the computation of real distances on the surface in terms of the coordinates. This mathematics, however, could do much more, and soon other mathematicians, including Hermann Grassmann and Bernhard Riemann worked out a very complete and general understanding of these "non-Euclidean" geometries that could apply to three-dimensional spaces (like that you perceive around you), and even four-dimensional space-time.[18] It came as a shock to the world of mathematics that curved spaces, in which parallel lines could meet, or in which the angles inside a triangle do not add up to 180 degrees, could make full and self-consistent sense. But they were generally considered to be quite abstract and strange curiosities, rather than having something to do with the real world.

Einstein was bold enough to imagine otherwise. If space-time is curved, just as the surface of the Earth is curved, and just as is described by the mathematics of Riemann, then gravity could be naturally and elegantly accounted for: the curvature would alter the space-time length along a path, and therefore change which path would be the longest and hence would be followed by an object. Because gravity is really a modification of the structure of space-time, rather than a force, *all objects would be affected in exactly the same way.* The "coincidence" discovered by Galileo that all objects fall at exactly the same rate is explained, and beautifully.

Einstein's prescription, therefore, was to suppose that in the absence of nongravitational forces, objects follow a path that maximizes their heartbeat time T. But T, instead of being calculated by the formula you used to CHOOSE YOUR PATH, is given by a similar but more complicated one that accounts for the curvature of space-time; only if space-time were uncurved would T be given by the simple formula. Such a path may not look at all straight when examined in space, or when drawn on a map. But secretly it's as straight— or more accurately as *long*—as it can be. To return to your leap off the cliff, the space-time path of the book, and of you, is straight. It is the space-time around you that is curved.

Following this logic leads to a radically new view of what it means to just sit still, and what gravity is. Consider a frame that is sitting at rest just off the edge of the cliff. Is it an inertial frame? No, it is not. Einstein's equivalence principle tells us that the truly inertial frame is one that is *freely falling*. But that freely falling frame is accelerating downward with respect to our cliff-edge frame. If we turn this around, our cliff-edge frame is accelerated with respect to an inertial frame, and so, in the cliff-edge frame we should feel "fictitious" forces. Indeed, we do! We feel a strange force, seemingly from nowhere, that grabs us and pulls us down: gravity! Gravity, in Einstein's theory, is a fictitious force just like—and no more or less real than—the centrifugal force on our arms as we pirouette or the forces inside an airplane passing through turbulence. This, really, is the way that space-time curvature manifests itself: inertial frames, in which space-time looks flat, are still present, but their relationship to one another is modified in interesting ways so that when we are in a gravitational field, *not falling* toward the body means we are *accelerating* with respect to the inertial frame, and we feel gravity. The gravity you feel pulling you down right now is *just like* the extra weight you feel in an ascending elevator—only it's a very big elevator!

What, though, determines how space-time is curved? Stuff does. That is, the more matter there is, the more space-time curves around it. The fact that Earth attracts you (and everything else) is equivalent to the fact that all of the matter in the Earth curves space-time in a particular way, so that the longest path, and hence the one that objects naturally follow, is one that veers toward the center of the Earth, rather than the one that looks like a straight line in space. To complete the picture, Einstein provided* a mathematical rule specifying the relation between the curvature of space-time and the presence of matter. Einstein's theory of general relativity could

---

* After, it should be said, many years of labor, of which he said: "Compared with this problem the original relativity is a child's play."[19]

reproduce all of the successes of Newton's theory of gravity, but it also predicted new and astonishing phenomena that had no previous explanation.

As you sit at the bottom of the cliff and take a moment to appreciate solid ground, you can perhaps see why Max Born called Einstein's successful formulation of his general theory of relativity the "greatest feat of human thinking about nature."[20]

But it's just a scratch in the mystery, and only a first step on our road.

PART 2

An Uncertain Trail through Treacherous Terrain

Penetrating so many secrets, we cease to believe in the unknowable. But there it sits nevertheless, calmly licking its chops.

—H. L. Mencken, *Minority Report*

10

RELEASING THE DJINN

(ARABIAN DESERT, 1610)

After endless, endless, endless sand, endless wandering, and boundless thirst, a cave. A cave! A cool, dark cave! It was sheer and amazing luck that you happened upon it, as hidden as it was. And you hear water! Just a trickle, but distinct. You stumble in, wet your tongue, then slowly drink your fill.

The cave is large inside, apparently uninhabited. But not entirely empty. There is a small, old, battered, brass lamp. Curious at the strange shine the metal gives off, you give it a brief rub. Who could resist?

Instantly and astonishingly, before you suddenly towers an immense creature—part human, part mist, fully daunting. Then the being speaks.

"I am the djinn Ibn-la-Plaz," it thunders, "and I am glad that you have finally arrived to free me from my lamp and cave. Freedom from the lamp has put me in a good humor. Tell me, would you like to know something of your future? For I see it, and see that you are curious."

It's hard to know what to think in these circumstances, but nothing comes to your addled mind beyond playing along. You reply: "Oh great djinn, I am most glad to be of assistance and grateful for your offer, but tell me first, by what sorcery do you know the future?"

The djinn nearly bellows: "Fool! It is the opposite of magic." Picking up the empty lamp, it continues: "If I drop this lamp, what will happen?"

"The lamp will fall, your greatness," you stammer.

"Yes! Did you use magic to predict that? No! Just like the lamp, every little bit of you follows an inviolable law that tells it what to do. By looking at you, I can instantly see where every atom will go, and thus what *you* will do. By looking at the whole Universe, I know everything that will ever happen. You will ask your question now, the question that is most important to you."

You think for a moment on what the djinn has said, and grow more and more troubled. Finally, you respond: "But what if I do not want to ask a question?"

With that, an enormous—albeit somewhat evil—smile spreads across the djinn's face, and it begins to laugh, louder and louder, until your whole world shakes.

Indeed the first thing that Allah created was the Pen. He said, "Write!" It said, "What should I write?" He said, "Write down everything that is going to happen." So at that moment flowed everything that would happen until the Day of Judgement. At this the Prophet was asked whether the actions we do are new actions or are actions that have been decreed already. He replied, "They are already decreed."

—Attributed to Muhammad by Al-Tirmidhī

We have thus far pieced together the basic components of a description of the physical world as developed over three centuries, starting with Galileo and culminating in Einstein's 1917 theory of general relativity. This understanding allows us to chart the path through space and time taken by a single idealized object—let's call it a *particle*—that moves under the influence of various forces. Such a particle's space-time path depends on three things: the structure of space-time itself, all of the nongravitational forces that the particle experiences, and the initial position and state of motion of the particle.

The structure of space-time determines which space-time paths are "straight"—that is, the maximal space-time distance T—even if they look curved in space (as does the path of the thrown book during its descent from the cliff) or as plotted on a flat surface (as for Kundulun Khan's maps of paths on the curved Earth); this effect is what we call the force of gravity. Nongravitational forces appear as deviations from these straight paths, just as we saw when looking at the effects of wind or magnetism on lead, wooden, or iron spheres rolling by the Leaning Tower of Pisa. Finally, the initial position of the particle determines the space-time point at which the path starts, and the velocity determines its initial "direction in space-time."

Given all of this information, we can compute the path very precisely. Take the movement of a spacecraft in the solar system. Its initial position and velocity are known very well by design. Because we also know very well the location of the Sun and the other large masses in the solar system, we can precisely compute the structure of space-time, and hence the gravitational effect on the craft; nongravitational forces are weak, but also fairly well understood. This is what allows us to direct missions through the solar system not by rocketing them about, but essentially by artfully choosing the right initial conditions for them so that they naturally end up just where we want them several years later!

In this picture, we've assumed that our spacecraft "particle" moves under the influences of *known* forces. Yet the space-time through which it moves, and the forces to which a particle is subject, are determined in turn by the presence and arrangement of *other* particles. The locations and masses of the particles determine how space-time curves, and other properties of the particles determine other forces, so to predict our spacecraft's path we also have to predict where all of the planets will be as well, and each of these predictions will depend on the others. Likewise, the nongravitational forces depend on the distribution of interplanetary gas, the brightness of the Sun, the presence of magnetic fields, and other such variables, which in turn have to be predicted. This might seem hopelessly complicated, and in practice it is indeed quite difficult. But the worldview of fundamental physics holds that underlying this complexity there is also an amazing *simplicity*.

The claim is that the seeming diversity of forces is largely an illusion. Essentially *all* of the nongravitational forces that we experience in the everyday world are really manifestations of electromagnetism, as described by Maxwell's equations. These equations govern how electric and magnetic fields are determined by charged particles, and in turn impose forces upon charge particles. The forces pertaining to particles and fields then underlie the forces with which you are intimately familiar. When a substance—solid, liquid, or gas—applies a force to you by pushing or rubbing, it can be meaningfully thought of as the electric forces between the atoms composing that substance and those composing you.

In this description, then, all of the relevant forces essentially boil down to gravity, caused by the distribution of matter, and electromagnetism, caused by the distribution of charges.* Thus, we can imagine the following procedure: We start with all of the particles that interest us, along with their locations and velocities. We also specify the fields that are present at that same initial time. A moment later, we update the particle positions according to their velocity. We also calculate the force on each particle that is due to the fields at its location; on this basis we update the velocity of each particle. Finally, we update the values of the fields everywhere in space by using the dynamics of those fields, as well as the locations and veloc-

* The strong and weak nuclear forces are important for the stability and behavior of atoms, but play little role at larger scales.

ities of all the particles. We can repeat this procedure for the following moment and, in this way, trace out in a self-consistent way the paths for all the particles.

It is hard to overstate the power and success of this procedure, using these theories. Newton's equations (or Einstein's, if necessary) make it possible to look up the dates and times (to the minute!) of eclipses centuries from now. The computer on which I wrote this book has billions of tiny circuit elements governed by Maxwell's laws.[1] Circuits can require intelligence, intuition, and patience to design so that they operate correctly. But there is no guesswork: physics tells you *exactly* what a given circuit will do.

On the basis of these successes, physicists had by the early twentieth century assembled a worldview consisting of *just three things*: space-time, particles, and fields. Particles give rise to fields, which inhabit and curve space-time. Space-time and the fields in it tell the particles how to move. This view forms a closed and self-consistent loop, and it is *predictive*: given the complete specification of the structure of space, the fields, and the locations of all particles, we have a mathematical procedure by which you can calculate the space, the fields, and the particles' locations at *any later time*. According to this picture of physics, if a being, such as the djinn Ibn-la-Plaz, were able to know at some instant the exact state of part or all of the Universe, it could then know the state at all earlier and later times.

Such a being is just a fiction; it would be impossible using any known or foreseeable technology to either determine or compute the paths of even a tiny fraction of the million billion billion (10^{24}) atoms making up even a tiny everyday object. The key point of the djinn here and in its various later manifestations is as a thought experiment regarding what is possible *in principle*, even if fiendishly difficult in practice. If the world were as described by early-twentieth-century classical physics, then *in principle*, all of the information necessary to predict the future (and past) with complete precision exists right now, even if we are not, in practice, able to make the prediction.

This line of reasoning seems to have been first coherently put forth in the context of physics by Pierre-Simon de Laplace (obviously named for the earlier fabled djinn) in 1821, and may be termed the philosophy of *scientific determinism*. As he elegantly put it:

> An intelligence that at a given instant was acquainted with all
> the forces by which nature is animated and with the state of the
> bodies of which it is composed would—if it were vast enough
> to submit these data to analysis—embrace in the same formula
> the movements of the largest bodies in the universe and those
> of the lightest atoms. Nothing would be uncertain for such an
> intelligence, and the future like the past would be present to
> its eyes.[2]

This view asserts the truth of a possibility we saw when BEING TIME: that at a fundamental level, a fateful decision that you will make tomorrow is *no different* from a decision you made yesterday. Both decisions are equally real, existing, and irrevocable; it just happens that the entity reading this sentence does not know about the decision made a day later. Nor is yesterday "gone" in any way: it is just as real and just as "present" as is the moment at which this next sentence is read. If the djinn existed and were sitting next to you, it could tell you what decision you would make tomorrow, and *nothing whatsoever you could ever do* could prevent you from making that decision. To the djinn, such an attempt would be as if one word on this page were struggling to change another one. As Einstein put it: "For us believing physicists, the distinction between past, present and future is only a stubborn illusion."[3]

This is a rather complicated chain of logic that ends up denying what seems intuitively extremely clear: that you can change the future (which does not yet exist) but not the past (which exists but you cannot return to), and that you really are free to make decisions that affect how the future will turn out. But whether this apparent freedom is *real* is a question that has long bedeviled those who think carefully about how the world is put together, in both religious and secular contexts.

A long-recognized *theological* paradox arises in trying to reconcile an omniscient God with our freedom to choose: if God (or Allah) is omniscient, then it (like the djinn or Laplace's intelligence) cannot be ignorant of what choice we will make, so the choice is already determined, if not already made. And from the religious and moral perspective, it gets even worse: If we do not have true agency, then how can we be held responsible for moral choices? What meaning do good and evil even *have* if our choices are not free?

While these puzzles manifest themselves almost inevitably with belief in an all-knowing deity, they arose long before any such idea was widely adopted. In Western thought, this idea was clearly in the air in ancient Greece with the atomism of Leucippus and Democritus, the naturalism of Anaxagoras, and the famous explanation by Socrates (to which we shall return) of the reasons for his staying in Athens to be executed.

In the ancient East, *karma*, which can be simply translated as "cause and effect," underlies a view of both physical and mental phenomena. It was a key part of the Buddha's thinking as well; he held that mental or physical occurrences are the inevitable result of the causes and conditions giving rise to them: "When this is, that is. From the arising of this comes the arising of that. When this isn't, that isn't. From the cessation of this comes the cessation of that."[4]

Scientific determinism represents, in a sense, a logical culmination of these ideas: there is a state of the Universe at some time, as well as fixed immutable laws, that together determine the state of the Universe at all earlier and later times. If we consider our thoughts and decisions to be part of that state, they are just as determined.

And the overwhelming success and predictive precision of this view undergird the claim that these laws do not brook exceptions; we are bound to their whims.

All of these arguments can be made at the level of Newtonian physics, as Laplace did. Does Einstein's relativity change things? As might be surmised from Einstein's statement that the distinction between past and future is a "persistent illusion," relativity leads to an even more powerful argument!

Suppose that determinism is false, and that there is a demarcation— the "present"—between the past, which is fixed, and the future, which is open and not yet decided. The Universe to the future of this "now" has yet to come into being.

But, "now" *according to whom*?

Special relativity teaches us that whatever definition of "now" we choose, there is another observer, in motion with respect to us, with their own definition of "now," that is equally valid. We saw that the planet being gobbled by the black hole might be to the future of "now" as determined by a given observer, but to the past as determined by another.

General relativity takes this ambiguity even a step farther. It holds that you can call *any* two points the "same time" as long as one cannot send a signal (traveling the speed of light or less) to the other. Thus, while we've previously shown "now" as a straight horizontal line across a space-time diagram, in general relativity, *any* curve can be "now," as long as it never increases more vertically than horizontally. This is depicted in the space-time diagram opposite, which illustrates different ways of slicing space-time into space and time. Each time is a continuous set of points representing events that are considered simultaneous—a particular COSMIC NOW. But simply requiring that simultaneous events cannot affect each other means that each "time" curve can bend up and down on the page. This flexibility allows many possible methods of defining time. In one method, the horizontal line labeled t_a might be a single time, just as in our previous space-time diagrams. But according to another definition, the curve t_b would be a single time. Events, like those labeled "1" and "2" in the figure, that are simultaneous in the first definition take place at different times in the second one. The curve t_c would *not* constitute a valid set of simultaneous events, because some of them are in causal contact—that is, could exchange signals.

This ambiguity in how to define simultaneity means that in either special or general relativity, some events that are occurring "now" according to your definition are to either the future or the past according to some other *equally valid* definition. To assert that there is a cosmic sense in which some parts of space-time are "already" fixed and determined, while others are "not yet," appears directly at odds with the principle of relativity.

These are powerful arguments suggesting that the future is predetermined. The arrow has already hit its target, or not. You don't know

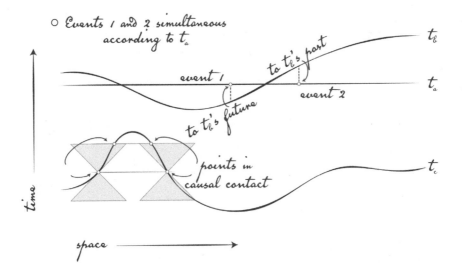

Different ways of slicing up the same *space-time* into *space* and *time*. (See text for explanation.)

which, but the event exists independent of your knowledge; you just have yet to get to it.

But are these powerful arguments *true*? Does this powerful djinn exist, if not as an actual being, then as a great mathematical Pen grinding away to determine all events throughout the Universe to eternity?

11

MANY PATHS MAKE THE ROAD

You awaken. Your eyes will barely open, and as you try to sit up, the immensity of your fatigue weighs upon you like a mountain. Where are you?

A small man sitting in the room's corner notices your awakening, hurries over to feel your brow, and smiles. He begins to speak in a foreign tongue that you do not immediately recognize, yet somehow you are able to piece together his meaning: you were very ill . . . high fever . . . not making sense . . . brought you from the road. Where *are* you? You manage to get this question across, and eventually you establish that you are in the city of Shenyang, heart of Liaoning and the domain of Kundulun Khan.

China?! The last thing you can clearly remember is a fantastical vision of a djinn; before that, your landing in the East, your abduction, your escape, and the desert route to Baghdad. After the desert, everything is a fog. But although you cannot remember the intervening journey, a vague mental map takes shape in your mind and leaves you dumbfounded by what you must have somehow crossed through, over what must have been months or years. But *how*?

Because you have apparently traveled so far, word of your predicament spreads, and you gradually gain help from several knowledgeable traders who have arrived from distant lands to the west. From these and the partial maps they provide, you assemble a view of all the possible paths that connect Baghdad to Shenyang. It is a daunting map. Which path did *you* take?

Some paths, obviously, are more probable, being far easier routes. But this still leaves many options, as well as the nagging possibility that you took a route that was much harder than it had to be. Some recovered memories help considerably: your recollection of the Helanshan rock carvings establishes that you skirted rather than delved deep into Mongol lands. Other memories are less helpful: your vision of an enormous blue lake would seem to easily pin down part of your route, but even with prolonged effort you cannot sort out which one it is—Adyar? Qinghai? Issyk Kul? Somewhere else? And there's a dove you keep remembering. What does that mean?

For a time, as the memories are slowly returning, you've created so many different stories for how you arrived that they start to take on a life of their own, and sometimes just as you are falling asleep they feel so real, as if you took *all* of those different routes.

An absurd thought, of course.

Thirty-one years ago [in 1949], Dick Feynman told me about his "sum over histories" version of quantum mechanics. "The electron does anything it likes," he said. "It just goes in any direction at any speed, forward or backward in time, however it likes, and then you add up the amplitudes and it gives you the wave-function." I said to him, "You're crazy." But he wasn't.

—Freeman Dyson[5]

In the worldview of classical physics we just discussed in RELEAS-ING THE DJINN, an object follows a path through space-time that depends on three things: the structure of space-time itself, all of the nongravitational forces that the object experiences, and the initial state of motion of the object. Let's look more closely at the last ingredient.

The initial state of motion is crucial because space-time structure and nongravitational forces determine a whole *set* of possible trajectories, while specifying the initial position and velocity of the object picks out a particular one. This is just like specifying a path on a piece of paper: if I tell you an ant walked across a piece of paper in a straight line, can you draw that path? Not quite. But if I provide you a point on the paper through which the ant passed, and a direction in which it did so, then you can draw the path with confidence.

Now consider an alternative. You could also draw a straight path for the ant if I labeled *two* spots on the paper through which the ant passed. Analogously, instead of choosing the initial position and velocity of an object's trajectory, we can instead pick the initial position and then also specify the position of the object at some later time. Together with the space-time structure and forces, this specification uniquely determines the space-time path of the object between these two times.

Thinking of a path as being determined by its beginning and end ties back to the last time we picked among paths, during the horseless descent from the mountain pass. There, the start and end positions and altitudes were fixed, and the issue at hand was how to descend from one altitude to another with the least effort. This was an analogy for the movements of particles under forces, where we saw that *forces* were mathematically equivalent to *effects that alter the "action" S*, and that *extremizing* the action (making it minimal or, as appropriate, maximal) was mathematically identical to saying that forces represent changes in velocity with time. And this method can be very elegantly combined with gravity. Rather than choosing the "straight" path as longest path through space-time, and then thinking of nongravitational forces as perturbing us away from that path, we can combine the steps: there is an action S that takes into account *both* the length of the space-time path and the other forces at play. With no other forces, extremizing this path gives the longest path through space-time, which manifests gravity if that space-time is curved in interesting ways because of matter. If other forces are present, they alter which path extremizes the action, so that it is no longer "straight" by our previous definition.

In short, given the start and end of a path, we have a mathematical prescription for determining exactly *which* path an object takes between them, even including gravity as curved space-time. This method is just as incredibly accurate as the method of calculating forces that push the object around. But while we're used to forces pushing us around personally, we're generally *not* used to thinking about our whole space-time path between two events. The foggy memories of a journey between the djinn's Arabian cave and a hut in China give us an opportunity to do so.

If, because of amnesia, you could not recall which path you actually took, we might try to reconstruct it for you, from whatever information we have. First, we could try to pick out all the possible ways you might have traveled between the endpoints, focusing on actual roads and known travel routes; a subset of these is shown in the figure on the next page. Without any further information, we would just have to guess among them, with each being no more or less likely than the next.

Suppose, though, that you can remember a particular and very interesting set of stone carvings that we are able to identify as existing near Yinchuan. In that case, tracing back westward from Shenyang (see the figure), we can conclude right away that you did *not* take the northern route through central Mongolia; you must have come from the southwest. But this still leaves many possibilities; among them are a due-westerly route that then splits into many paths criss-crossing Mongolia, or a much more southern route through Tibet and India. Which did you take? If we could determine whether a lake you can vaguely recall was Qinghai Lake in Tibet rather than

Possible routes between the djinn's cave and Shenyang. Tracing west from
Shenyang, the probabilities shown assume a 75% = 3/4 chance of having
passed by Qinghai Lake, none for the most northerly route, and equal
probability assigned to either possibility at each branch heading west.

Adyar Lake in (what is now) Kyrgyzstan, we could further narrow
down which route you actually traversed, just as when we used the
stone carvings.

But what if you don't know for sure? In that case, we might still
assign *probabilities*. Suppose you were 75% = 3/4 sure that the lake
was Qinghai, but you accord a 25% = 1/4 chance that it was, in fact,
Adyar. Then, as labeled in the figure, we would assign a 3/4 chance
to the most southern path in the figure, whereas the tangle of more
northern paths would all share 1/4 total (except for the most northern one, which gets near-zero probability because it does not pass
through Yinchuan). If we wanted to determine the probability for a
given northern path, we might, for example, assign equal likelihood
to each tine at each fork in the road encountered as we reconstructed
your route back from Shenyang. Thus, for example, we would accord
a 1/8 chance to each of the two routes headed around the Taklimakan desert as we trace back west from the Qinghai-Adyar split.
By splitting (and recombining) probabilities, we would end up with
a set of numbers like that in the figure.

We now seem to have two quite different ways to assess paths between two locations. In one, there is a single, particular path that we can determine by extremizing the total action S of the path. In the other, we accord each path a probability; in the absence of any information, we can give them all the same probability, and if we gain some information about the path taken, we alter the probabilities to reflect that information.[6]

These sound like two completely different ends of the spectrum: on one end all paths are "equal," and on the other end there is just one "true" path, and that path actually gets followed.

What Richard Feynman did was to show how, in a very deep sense, and in a way that will crucially address the issues posed by the djinn, these two ends of the spectrum are *exactly the same*: if you follow all paths equally, you end up just following a single path. The one true path.

As Freeman Dyson said: That sounds crazy. But it isn't.

12

SUFFICIENT REASON FOR A ROLL OF THE DICE

While technically forbidden, gambling was still loosely tolerated in the court of Jahangir, and your recollections of Cardano's unpublished book on games of chance makes you feel that you should have the edge in games of dice. However, luck has not been with you, and you've started to complain loudly regarding your poor rolls. The stakes have gotten high.

A passing holy man overhears and chides you: "Do not complain. Your losses at dice are the will of Allah." Given the state of your finances, you do not find this commentary helpful, but you reflect on the Sufi's words. What *does* cause the roll of a die? The number that comes up does not appear to be truly random, but rather the result of innumerable and complex causes—the angle of your hand, the exact speed imparted, the texture of the floor, and so on—that conspire to make the roll difficult to predict. Your time with the djinn, however, has largely convinced you that, while difficult, predicting the roll of a die is not impossible.

Staring at the dice in your hand, you feel that for you, there is 1 chance in 3 of a favorable outcome. But for the Universe—or Allah, or the djinn—the die is as good as cast, and there is 100% probability of a particular outcome. If only you could know what it is! You console yourself with the fact that your gambling companions are just as limited, and that you are not gambling with the djinn, which would be a hopeless prospect. You begin to wonder, though: Could there be a die that not even the djinn could predict? Perhaps. But what about God? Or the Universe? Could there be a chain of causation that starts with no cause?

   You pull your mind back into the game, and prepare to roll the dice . . .

Can there be an event without a cause? We're certainly used to dealing with events that we cannot predict—like the roll of a die, or next week's weather. We're also accustomed to the idea that while unpredictable in a particular instance, many events are *statistically* predictable: the chance of rolling two *6*'s on a fair pair of dice is 1/36, or 2.78%. By "fair" we mean that for each die, the chances of landing on any number from *1* through *6* are all equal. By "chance," we generally mean that if we imagined rolling two dice many, many times and counted up the results, we would find that about 1 in 36 rolls has two *6*'s. (This is the sort of regularity over time that allows casinos to steadily get richer!) Beginning with Girolamo Cardano and developed through the seventeenth and eighteenth centuries by Laplace, Descartes, and others, there is a well-developed mathematical theory of probability that can be used to assign probabilities to events that we cannot definitively predict.

   We use these theories because we don't know which way the die will land or, say, whether it will rain tomorrow. But does the *Universe* know? It's perfectly clear that there are *reasons* a certain die face comes up, or that it rains tomorrow. Supercomputers at this moment are working to predict tomorrow's weather much better than you could. Or consider a cast die. A short video clip of the die just thrown, combined with a sophisticated physical computer simulation of the die, table, air, and so on, would be enough to predict the die's landing with very high accuracy. It would be foolish to bet against such a system if it were allowed to call the result during flight!

The computer's ability to predict how a die will land raises two questions. First, what happened to our 1-in-6 chance for a given side to show on the die? Clearly, this is not how the computer sees it; it instead assigns much higher probability to one side than to the others. Second, where did the 1-in-6 chance come from in the first place?

Let's look more closely at what the fancy computer system—call it a *simulator*—would do. The short video clip would constitute a set of measurements, including the position and velocity of particular parts of the die. The simulator would then use these as initial data, numerically solve basic physics equations, and output the result of the simulation, indicating that (say) the *4* side would be face up.

But a well-designed simulator would not stop there, because even if the simulation were perfect, the measurements provided by the video clip aren't: each measured number has an associated uncertainty. A good simulation, rather than running only once, would run many, many, many times, each time using a slightly different set of initial conditions chosen from the full range of possibilities for each measurement. For example, if the initial speed of the topmost corner of the die were measured from the video clip to be between 4.5 and 4.7 cm/s, simulations might be run in which this speed was assumed to be each of 4.50, 4.51, . . . 4.70 cm/s. And for each of these 21 values, other measurements (like the initial position, or direction) could be varied. The result would be lots of simulations, potentially with lots of different outcomes.

Now, the simulator can tally up the fraction of times that a particular number (*1*, *2*, etc.) appeared in this whole suite of simulations. Those fractions translate into a good set of probabilities to assign to the outcome of the throw. The simulator might output, for example: "Out of 100,000 simulated throws with initial conditions consistent with the video clip, 3% came up *1*, 96% came up *4*, and 1% came up another number." This is an extremely useful prediction, giving not just the most likely outcome, but how *much* more likely that outcome is than others. This is also exactly what weather simulations do: the same simulation is run many, many times, and the fraction of those simulations showing it to be raining in your location tomorrow is quoted as the "chance of rain."

Now let's return to the 1/6 probability that we, lacking the simulator's predictive power, accord the result of *4*. That probability has a somewhat different origin. We could imagine throwing the dice many times and counting up results, but there's no self-evident reason to get 1/6 from this exercise. Rather, the probability arises from the *symmetry* of the die: other than the painted numbers, the sides of the die are operationally the same. More specifically, there is no *correlation* between any particular side of the die and the physical processes involved in the die's rolling; this is really what we mean by a "fair" die. If the die were weighted more on one side, it would be "unfair" precisely because there are correlations between a particular side and the physical processes involved in a throw, and these correlations break the symmetry of the six sides.

Symmetry, though, is not quite enough to explain the difference between "your" 1 in 6 and the computer's 96%. If you drop a die from less than 1 cm above the tabletop, it is quite likely that it will land with the same side up as when it left your hand. So, the 1-in-6 chance requires not just a lack of correlation between the side of the die and the physics governing its motions but also sufficient complexity in the physical process to break any correlation between the outcome and the sort of information to which you as a human die roller have access. That is, links between initial conditions and the outcome are present in a regular throw of the dice, but it takes the observational and computational power of the computer simulator to both see and use those links to derive probabilities that differ from 1 in 6.

Let's sum up. During a throw of the dice, both you and the computer simulator go through a very similar predictive process. You have a model for the throwing process, as well as access to some set of information about the particular throw you are predicting. For you, this information is pretty useless, and you resort to an estimate based on symmetry to give equal chances to each side. The simulator, which has access to useful information and the capability to employ a much more sophisticated physical model, can make a more predictive probability distribution that would allow it, for example, to win money from you in betting.

It's easy to imagine the simulator being better or worse at its job. A better video camera, a better physical model of the table and the die's bounce on it, more computing power, and so on, could all "concentrate" the probabilities for a normal die throw so that, for example, 4 became 99.6% probable rather than 96%. But it's equally easy to imagine the process going less smoothly; for example, in a throw where the die lands on an edge, the probability might be near 50-50, and it might take a great deal of simulation refinement before the probabilities start to fall one way or the other. Or if the die were rolled down a long, bumpy hill, the simulator would be hard-pressed to do better than the standard 1-in-6 chance, because even it would be confounded by all of the variables and uncertainties. Still, it seems that with tons of effort put into refining the model and collecting better data, the simulator should eventually converge on a single highly probably answer as to what will happen.

Where, then, is the limit? Is it possible for a "predictor," with capabilities like the djinn, to be so good at its job that it always accords 100% probability to one particular outcome, and zero to the others? In this case, you could say that you know *all of the reasons* behind the outcome of the dice throw, and have fully answered the question of why a particular side came up.

Although this seems hard, it is not obvious why it should be impossible *in principle*: even if we don't have the necessary information and know-how to predict what will happen, it seems intuitively clear that the information and know-how *exists*, since the

Universe has the information and makes some particular thing happen rather than something else. There should be a reason for that, right? The great Enlightenment philosophers Descartes, Spinoza, and Leibniz disagreed on many things, but they agreed on this: there is always a *reason* why one thing happens rather than another thing. As Leibniz put it in his *Monadology*: "No fact can be real or existing, no statement true, without a sufficient reason for its being so and not otherwise."[7] It may be enormously difficult, but a perfect prediction—or a perfect explanation—should always be possible.

We often say something happened "for no reason," but we rarely actually mean this. Can you really conceive of an effect without a cause? Some very great minds have tried, failed, and decided that it isn't possible.

And yet, according to what is probably our most fundamental theory of physics—quantum mechanics—it is.

... As you are about to roll the dice, you reflect on how simple a thing it is—a single roll of the dice—that could so greatly affect your fortune and fate. Just six possible numbers on each die, nothing in between and no ambiguity, paring away strategy, skill, history, and the world's complexity into pure simplicity.

You roll, and your course is set.

Imagine a die that, rather than being made of an enormous number of particles, *is* an elementary particle. In particular, try to imagine that the die has only *two properties*: what side (*1–6*) is facing up, and where the die is. Each of these properties corresponds to a question you might ask of the die—in other words, a measurement you might make. We might have, for example:

Question 1 (Q1): Which of your six sides is facing up?

Question 2 (Q2): Where exactly are you?

And the die might give these answers:

Answer 1 (A1): My *6* side is up.

Answer 2 (A2): My [latitude, longitude, altitude] is
   [27.1789335252, 78.0224962785, 1.232432].

This idea leads us to an important and far-reaching definition. We'll call the *quantum state* of a physical system a *complete set of definite facts about itself that the system would provide if asked.* The quantum state of our die, then, would be composed of the two answers A1 and A2, combined in shorthand as [A1 ; A2] = [6 ↑ ; 27.1789335252, 78.0224962785, 1.232432], with the semicolon separating the different questions being answered.

What is crucial is that these answers are *all the definite information that the die has to give.* That seems fairly innocuous, but be warned: from this statement of the fundamental simplicity of a system, a whole lot of counterintuitive results will follow. Let's see how.

First, it's clear that there are different states our die might be in. To describe these, we can imagine making a *complete* (all possibilities are included) and *mutually exclusive* (only one can be true of the system at any given time) list of answers to each question. For the die, this would include six possible sides up in the A1 slot, and all possible positions of the die in the A2 slot. The quantum state of the die might correspond to any one of these possible sets of answers, and all possible definite answers the die can give are somewhere in the set.

Now we get to the crucial point. Although we've listed just two questions, what if we nonetheless go and ask a different question, like Q3: Which side is facing east? We then have a conundrum: Q3 is a valid question, corresponding to an experiment we could actually do. We can look from the east and see what we see. Thus, the die *must* provide an answer.

So it does. But the answer can't be *definite*, now can it? We have an exhaustive list of questions to which we get definite answers, and Q3 is not on it! Thus, it must be the case that the answer *might* be *2* or *3* or *4*. (It might even be *6*, even in a case where we know that the *6* side is facing up, not east!) There is unavoidable uncertainty as to which answer the die will give.

This does not mean that all possibilities are equally likely. Quantum mechanics gives a very clear mathematical prescription (called *Born's rule*) for how likely each answer is for a given state of the die; that is, it gives a *probability* for each answer before we make the measurement.

And thus have probabilities entered into what seemed like a straightforward question about a system *about which we know everything there is to be known.*

It is very, very hard for us to imagine physical things that are fundamentally simple in this way. Imagining our truly simple quantum die on a table with *6* facing up, one also irresistibly imagines it sitting there, with (say) the *4* side facing east, *2* facing south, and so on. Yet that's not right! Sitting still would be a property of zero velocity, and 4 facing east would also be a property. The die, though, has *only* the two properties of position and which side is facing up. This limitation strongly clashes with our intuition. When we define a property that some object can have, it's easy to forget that we *invented* that property, because normally, as soon as we invent a property, it's generally safe to imagine that the object either *has* that property or *doesn't have it*. And with that squared away, there seems to be no obstacle to moving on to another property and another, with no obvious limit to how many we can think up. Yet quantum reality is not this way.

THERE'S A BEAUTIFUL and precise way to describe all this in quantum mechanics, called *superposition*. We can think of a superposition of states as a weighted set of answers to one question in terms of answers to *another* question. For example, in a quantum die,[8] state [5↑], one answer to Q1, is the very same state as a sum over states that would give precise answers to Q3 regarding which number is facing east. This is represented as:

$$[5\uparrow] = C_1[1{\rightarrow}] + C_2[2{\rightarrow}] + C_3[3{\rightarrow}] + C_4[4{\rightarrow}] + C_5[5{\rightarrow}] + C_6[6{\rightarrow}],$$

where the right-pointing horizontal arrows signify the easterly direction, and $C_1 \ldots C_6$ are numbers. This expression indicates that the quantum state [5↑] gives a definite answer to Q1 but contains all six possible answers to Q3; the probabilities are obtained from the numbers $C_1 \ldots C_6$, which describe how *much* of each of the states [1→] . . . [6→] there is in [5↑].[9] The probability of measuring [3→], for example, turns out to be 1/16. Superposition is thus another way to say that a property is indefinite: a system has a little bit of one property, and a little bit of another property. But only one of those possibilities reveals itself upon measurement.

What about after the measurement? Well, at that point we know the answer to question Q3, since we just found it out. Then the state of the die *must* be one with a definite answer to question Q3. That part of the state might be [3→] if we measured that value, and the whole state could be, for example, [A3 ; A2] = [3→ ; 27.1789335252, 78.0224962785, 1.232432]. We've thus determined that if we ask a question to which the system is ready to provide a definite answer, then we get that answer, and we've done nothing to the system (other than learned something about it). But if we ask a question that the system is *not* prepared to answer definitely, we still get an answer, and in doing so the system assumes a state in which that question has a definite answer: the one we just got.[10]

There's just one more ingredient we need to complete the picture. Suppose we ask the system some questions and get some answers, so that the system is then in some state corresponding to the answers we just got. Now let's leave it alone. What happens to it? Quantum mechanics also says that, of its own accord, the quantum state changes in time according to a specific equation, which is named for Erwin Schrödinger. Analogously to Newton's laws and Maxwell's equations, Schrödinger's equation takes the *initial* conditions for a system—its state—and tells you its state at all later times. It therefore provides a description, and a prediction for what the system will do, in terms of how its state evolves.

THESE IDEAS—the quantum state, superpositions, Schrödinger's equation, and probabilities for measurements—are a lot to take in at once, and they take some getting used to. The good news, though, is that this is just about *all* there really is to quantum theory. While there is an incredible amount of subtlety in the accompanying ideas and implications, and a lot of technical machinery in applying them to specific systems, the core of the theory has basically just these few ingredients.

Let's then sum up, comparing the quantum formalism to how our (classical) simulator would make predictions. To start, we define a particular measurement process for determining which side of the die is up, along with a set of states, each corresponding to one of the six possible outcomes of the measurement. Next, we assign a state to the system at a given initial time, based on knowledge or measurements of the system. We evolve this state using Schrödinger's equation. Finally, we ask our question: Which side is up? Born's rule then takes the state at the time of measurement, compares it with each of the outcome states, and on the basis of this comparison assigns a probability to each outcome.

We thus have a pretty clear analogy between how on one hand quantum mechanics, and on another hand a "simulator" based on nonquantum physics, would predict die rolls by evolving some initial conditions for the die's roll (see the figure below). Both processes would lead to uncertainties in the outcome. Those uncertainties, though, are fundamentally different. The simulator is uncertain because it is slightly uncertain of both the initial conditions and the dynamics, and those minor uncertainties grow into major ones through the die's evolution. We imagine, though, that we could improve this situation by getting better cameras, faster computers, and cleverer programming.

In the quantum case, even if the initial state were *perfectly known*, it would not necessarily lead to a definite answer to a question of interest, even if we asked it immediately. Worse, the dynamics of

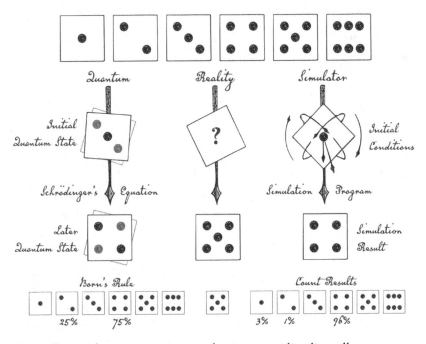

Using the simulator, or quantum mechanics, to predict dice rolls.

the system make it extremely unlikely that the die's state will provide a definite answer to the question we ask it later; as we'll soon
see in more detail, Schrödinger's equation tends to "spread out"
states of precise position into progressively less certain position or
velocity. To be clear, quantum uncertainties in everyday objects are
extremely small, and you could go a long way in making a ter-
rific dice-throw simulator without having to worry about them. But
they would appear at some level of accuracy, and be completely
unavoidable.

So, can there be rolls of the dice that even the djinn—who has
perfect understanding of the world—could not predict with certainty? Quantum theory says yes: in attaining its perfect knowledge, the djinn obtains a state for which the predictions of some
observables are necessarily probabilistic rather than definite. Even
knowing everything there is to know, and knowing exactly how the
state corresponding to that knowledge evolves, the djinn has no sure
answer to certain perfectly valid, important questions. And there is
no definable sufficient *reason* the djinn could give as to why one outcome occurs rather than another.

We still might object: when the die is cast, doesn't the *universe* know
what will happen? Even if we can't know the answer, if we can ask a
well-defined question, doesn't the answer exist?

A question cannot be answered until it is asked, and once asked, it
cannot be unasked.

13

THROUGH THE GATES

(SAMYE MONASTERY, TIBET, 1612)

From the hilltop you have an excellent view of the monastery, as well as an escape from the festival crowds. For several hours you have been watching pilgrims file through a narrow gateway to the temple complex. Each pilgrim stepping through starts to turn a handheld prayer wheel, and you are amazed to see that each turns the wheel at just the same rate—one revolution per step—while meditatively walking, all at the same rate, in the general direction of the monastery where they are gathering across a wide pavilion.

The traffic has been steadily increasing, so that pilgrims are lining up at the entrance, and at some point a monk scurries out of the monastery to open a second gate just to the north of the first, to relieve the congestion. Very soon there are two steady streams of pilgrims coming through at such a rate that, focused on their prayers, they are nearly running into each other.

You are amused to note that when two pilgrims approach each other and their prayer wheels are synchronized, they are able to avoid each other. But if one's wheel is pointing north while the other's points south, the weighted strings used to spin the wheels get tangled up, and the two pilgrims are forced to stop and sort things out.

Then you notice a fascinating phenomenon. Along the edge of the pavilion there are clusters of collided, disentangling pilgrims, separated by bands in which the pilgrims are passing through freely. As you watch, you see that this phenomenon persists through time: for some reason, particular zones of the pavilion edge lead to many collisions; past these are regions with few pilgrims stepping onto the pavilion. Other regions lead to very few collisions and many pilgrims stepping on. Something nags at the back of your mind: this reminds you of observations you did with Galileo back at the docks, but you can't quite make the link. Frustrated, you decide you've been in the high-altitude sun long enough, and you make your way downhill to the village.

The next day you return to your spot. A few straggling pilgrims are still wandering in, one every couple of minutes, and you see that the monk has forgotten to close the north gate. But as you sit there and watch the stragglers, the oddest thing slowly dawns on you. As the pilgrims approach the pavilion's edge, just like the day before, many of them step onto the edge in some places, and other regions appear to be avoided. This seems impossible, so you start to keep careful track and verify that exactly the same pattern played out by the crowd the day before is being recapitulated by the stragglers, even though none of them ever run into each other.

How can that be?

The gateless gate lies open.

Dharma gates are endless. I vow to enter them all.
                                        —The Bodhisattva vows

Is there a "sufficient" reason you took a path through Tibet rather than Mongolia?

Let's descend from the hilltop perch above Samye Monastery, and look closely at the two gates leading in. An individual might have to choose one gate or the other, but a pair of pilgrims can go through both at once. What happens when they do?

Imagine that, as depicted in the figure below, two pilgrims enter the north and south gates at about the same time, and as they step through, both of their prayer wheels have the spinner pointed east. As they walk, at exactly the same speed, the wheels spin one time around for each step they take. The pilgrims, because of their initial directions, might converge at the edge of the pavilion after exactly 20 steps, at a location due east of the point between the two gates (bottom set of paths in the figure). Since their prayer wheels have each undergone 20 full rotations, the spinners are both pointed due east, and they do not interfere with each other.

But now imagine (as in the top set of paths in the figure) that two pilgrims meet a bit more to the north. Then the pilgrim from the south gate had to walk 20¼ steps, ¼ step farther. In this case, the extra quarter rotation of the southern pilgrim's wheel means that the spinner is pointed *north*. Unfortunately, the northern pilgrim, having gone only 19¾ steps, has his spinner pointing *south*. Thus the spinners get all mucked up, and the pilgrims are unable to move forward.

There's a nice mathematical way to describe such pilgrims, using *complex numbers*. For these purposes we can think of a complex

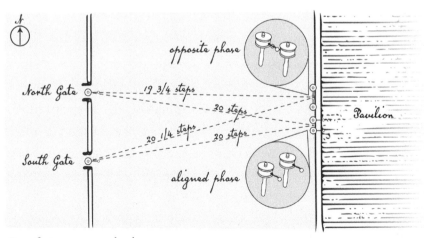

Interfering prayer wheels.

number as just a little arrow on a plane that, like any arrow, has a length (or *magnitude*) and a direction (or *phase*) between 0 and 360 degrees. A crucial thing about complex numbers is that when you combine them, you must worry about the magnitude *and* the phase, just as with a real number you must worry about the magnitude of the number and whether it is positive or negative. But the procedure is not that complicated. To add two complex numbers represented by arrows, you put the tip of one arrow at the base of the other. An arrow heading from the base of the first to the tip of the second represents the sum.[11] The length of the resulting arrow (and hence the magnitude of the sum) may be as long as the sum of the lengths of the two input arrows, or as short as the difference in their lengths.

The pilgrims' prayer-wheel spinners are just like this, except that we have given each spinner the same length, so all that matters is the direction of the spinner, which steadily revolves through 360 degrees each time the pilgrim takes a step. If two pilgrims come to the same place, we can think of their interaction as an addition between the complex numbers represented by their prayer-wheel spinners. If the spinners point in opposite directions, they cancel each other out (i.e., the sum has zero magnitude), while spinners pointing in the same direction represent complex numbers that nicely add (with the sum having double the magnitude). For other orientations, the magnitude of the sum might be longer or shorter than for the initial two arrows, and in various directions.

Such interactions, when evaluated along the edge of the pavilion, give some areas (like the center) where magnitude is high and there are lots of pilgrims stepping onto the pavilion; in others (such as just to the north), the pilgrims are all tangled up and few step onto the pavilion. In between, the number of pilgrims smoothly varies from high to low in correspondence with the directions partially canceling, then partially reinforcing. This phenomenon, known as *inter-*

ference, is a common property of *waves*. Waves, after all, also carry a magnitude (how high/low the wave is) and a phase (peak or trough), and can indeed be described by just the same mathematics.

Interference between waves was known to physicists even in the seventeenth and eighteenth centuries, and it was used to try to settle a rather bitter dispute as to whether visible light is fundamentally a particle like a mote of dust, or a wave like a water wave. A famous experiment, first performed by the British physicist Thomas Young in the early 1800s, shows that if you pass light through two slits to then illuminate a screen, the illumination has darker and brighter bands, just as it would if light were a wave with light from one slit interfering with light from the other.

But this victory for the wave view was transitory. In the beginning of the twentieth century, pioneering work by Einstein (for which he was awarded a Nobel prize) showed that light *does*, in fact, have clear particle properties, in that it exists in discrete individual packets— photons. How does this square with Young's wave experiments? Simply by being baffling. One can run the two-slit experiment by dialing down the intensity of the light, and putting in a really sensitive detector that registers a single photon at a time (This is just like coming back the day after the festival to find only one pilgrim coming through the gate every now and then.) Each photon lands at a well-defined location, just as a particle (or a pilgrim) should. You would think, then, that each of these particles would go through one slit *or* the other. Yet looking at the pattern of light particles, they also conform to the dark and bright bands seen previously: more particles in the "light" areas, and less in the "dark" ones. So the photons are interfering like *waves*, even though they are *particles* going through the slits one at a time. Even if you imagine that a photon goes through a slit, it somehow "knows" whether the other slit is open. This behavior was extremely baffling to the physicists of the early twentieth century.

———————

TO MAKE SENSE of things, quantum theory's developers formulated a mathematical representation called the *wavefunction*, which is
closely related to the *quantum state* discussed in SUFFICIENT REASON
FOR A ROLL OF THE DICE. Recall that we can write the quantum
state of something as a sum (or superposition) of states that would
correspond to definite answers to some set of questions, like "where
are you?" *or* definite answers to a complementary question, like "how
fast are you moving?" but often not both sets of questions at once.[12]
We can think of the wavefunction of a photon in our beam as the
photon's quantum state, written as a sum of *states of definite location*.
It thus directly gives probabilities for the photon's location at a given
time. The wavefunction at the barrier holding the two slits might be

[Wavefunction at slits] = [at northern gate] + [at southern gate],

so a measurement would have a 50-50 chance of finding the photon
at each of the gates.

　　Quantum theory also gives a well-defined procedure to *calculate* the wavefunction on the screen: just as Schrödinger's equation
can—as we saw in SUFFICIENT REASON FOR A ROLL OF THE
DICE—evolve the quantum state, it can also evolve the wavefunction from each slit to compute it at the screen's location. Because
the wavefunction is precise in terms of the position near each slit,
it turns out to be quite *imprecise* in terms of velocity. This means
that the location of the particle tends to spread out, much as a wave
emanating from a particular point spreads in all directions. These
waves/wavefunctions emanate from each slit and then, crucially, you
add the two wavefunctions from the two slits to get the combined
wavefunction at the screen.[13] Since each wavefunction has a "wavy"
nature (i.e., both a magnitude and a phase), wavefunctions can interfere, leading to the interference pattern composed of dark and light
spots where the probability of finding a photon according to the
wavefunction is low or high.

So, photons are particles if you ask where they are, but if you ask where they *will* hit the screen, the answer comes in terms of waves. (Waves of what? Probability!) But *after* they hit the screen, they are particles again.

There is a long list of perplexing and interesting questions to ask about this state of affairs, and physicists have kept busy asking them for a century now. Does the photon "really" go through one slit or through two? Can we figure out which slit the photon went through after it hits the screen? What about before it hits? What if we don't look at the screen? What if you close one slit after the photon goes through but before it hits the screen? And so on.

BUT RATHER than tackling any of these, fun as it would be, let's return to Feynman's perplexing statement that "the electron . . . just goes in any direction at any speed . . . however it likes, and then you add up the amplitudes and it gives you the wave-function." If we consider particles (or pilgrims) individually, there are many, many possible paths they might, *in principle*, take between one place (such as a gate) and another (the edge of the pavilion) later. Classical physics selects just one of these paths as the *right* one. Quantum mechanics undermines this idea: if a particle is at a gate, then Schrödinger's equation can tell us only the *probability* of a particle being someplace at the pavilion edge later. We might think of this probability as applying to *which* path the particle took, but that's not quite right: we saw from the pair of pilgrims that if there is to be interference, the particle has to—somehow—take *more than one* path.

Feynman took this idea to its logical conclusion. He asked: If the particle must take more than one path, what if it takes *all possible* paths? In his analysis he used an ingenious argument. Imagine adding a barrier between the gates and the pavilion, with a single gate inside that barrier. Then whatever paths the pilgrims took, those paths would have to pass through the gate in that barrier. If we added more gates, and more barriers, we would enforce a certain set of constraints on paths taken by the pilgrims. Quantum mechanics can be used to compute the probability of each possible path through the gates, just as it was used to compute the probability of a pilgrim at each position on the edge of the pavilion. What Feynman, amazingly, showed, is that if you consider an *infinite* number of barriers, each covered by an *infinite* number of gates, then the probabilities you get can be described in two ways.

First, you can say that they are a sort of sum over *all possible paths* that the pilgrim could take to the pavilion, since for every path you might choose, there is a set of gates to go through.

At the same time, a barrier completely filled with gates is no barrier at all—a barrierless barrier! Thus, the probabilities *also* describe just the "free" movement of a pilgrim to the pavilion, as described by the wavefunction with no barriers at all (see the figure below).

So, the wavefunction that describes a single particle traveling from one place to another is mathematically equivalent to a particle fol-

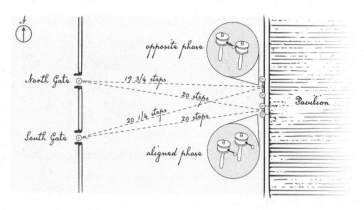

Gateless gates and barrierless barriers. The pavilion is marked by a stupa. Which path to take?

lowing *all possible paths* from one place to another, with absolutely no discrimination between them. It's crazy, as Dyson said, but it works!

It also leads to some pretty puzzling implications. One crucial point is that for the method to work, it is necessary to attribute *equal magnitude* to each and every one of the possible paths. No paths—not the straight path, or the path dictated by classical physics, or any others—are inherently privileged over others, nor is it meaningful to say that the particle took one path versus another. The paths give us the wavefunction, which gives us the probabilities, which give us (uncertain) answers to questions, like where we ended up.

And yet, objects follow straight-line paths, determined by the laws. And yet, when we ask how we got here, we took a path, which we remember.

You are made of particles, which take all possible paths. How can you take just one path?

14

SPLITTING THE WORLD

(EDO, JAPAN, 1624)

In the darkening night, Munenori is watching his opponent's eyes, left Achilles tendon, and sword, measuring the deflection angle of the sword by focusing on a half-centimeter glint of light near the hand guard. The slightest movement of the blade betrays itself by a momentary brightening of the glint: 958 extra photons reflect from the blade and hit Munenori's right eye.

Of the photons hitting his eye, 832 are absorbed in the eye before hitting its retina. Of the remaining 126 photons, 87 impinge upon 75 different rods; most of the rest are absorbed by cones but do not trigger a response. In 70 of the rods, special molecules absorb the photons, change shape, and cause a chemical response sending a signal through several levels of nerve cells, which triggers an enhanced response in 34 cells that connect to fibers of Munenori's right optic nerve. These fibers, in turn, lead to Munenori's visual cortex. The information they carry is combined in an incredibly complex manner through several additional neural systems, eventually leading to a split-second assessment by Munenori to expect an upward thrust of his opponent's sword. His analysis is correct. Munenori parries the blow deftly and, with a quick counterstroke, dispatches the enemy assassin.

Or does he? The glint of light from the sword *also* has 959 *and* 957 photons in it; it is *also* true that 124, 125, 127, *and* 128 of these hit the retina and are absorbed by 69, 70, 71, 72, 73, 74, 76, *and* 77 rods. These lead to 31, 32, 33, 34, 35, *and*

36 signals to the primary visual cortex. However, the groups of 31, 32, and 33 signals are insufficient to be noticed during Munenori's split-second decision process, and he instead prepares for a downward thrust. This decision, unfortunately, proves quite fatal.

Does Munenori live, or die, or both? And you, who have cast your lot with his—what do you do?

> We could go still further, tracing the chemical reactions which produce the impression of this image on the retina, in the nerve tract and in the brain, and then in the end say: these chemical changes of his brain cells are perceived by the observer.
>
> —John von Neumann, *Mathematical Foundations of Quantum Mechanics*

The Tibetan pilgrims have taught us that the quantum state of a system is very peculiar. It tells us probabilities for things to "happen," but not in quite the way our simulations told us the probabilities of outcomes at dice. Along with inevitable uncertainty, the quantum state can tie together outcomes in a different way than we're accustomed to in everyday objects. We might describe the probabilities for a die to land one way or another, but generally we consider those *separable* from what some other die might be doing. Not so with photons: we saw that the results of photons passing through two possible slits are not just the results of photons passing through one slit *or* the other; there is a real sense in which each photon passes through both slits. Thus a photon, even while being quite tiny (as we discover by its effect on a detector), is somehow also a very big and holistic thing, spanning the much larger distance between the two slits.

These peculiar features of the quantum state invite an interesting and important question: Why do we use quantum mechanics when describing the two-slit experiment with photons, but (Tibetan pilgrims excepted) classical mechanics and probabilities when describing things like rolling dice? The photons traveling from the sword to Yagyū Munenori's eye clearly should be treated quantum mechanically, as should the chemical processes in the eye's photoreceptors. What about the optical neurons? What, then, about the neurons in the brain?

There is a pragmatic question here: Quantum mechanics was devised because classical physics failed to describe some systems correctly. For what sort of systems is this the case, and for what systems can we do without quantum mechanics? There's also a more foundational question: Are there systems that *are* quantum mechanical, and others that *are* classical? If a photon in some sense goes through both slits, is there a real sense in which a monk goes through two different gates at once, or in which Munenori both parries *and* is cut down by his opponent's sword? Or are monks and samurai *really* classical things, for which we use quantum mechanics to make predictions? Let's tackle these by looking in more detail at what happens to the photons entering Munenori's eye, according to quantum theory.

WE BEGIN with the glint on the sword. The glint is composed of a number of photons, but this being quantum mechanics, this number is necessarily somewhat uncertain. Therefore, the quantum state of the glint contains a *superposition* of states, corresponding to a *range* of different numbers of photons. For illustration, let's say there are just two terms, one with 957 photons and one with 959, and we'll call them *[957]* and *[959]*. So, the glint quantum state would be a superposition of these two states. If we use font sizes to represent the magnitudes of states (their length, if we think of them as little arrows), then the state of the glint when it leaves the blade could be taken to be:

$$[Glint\ state] = [957] + [959].$$

The slightly larger font for [957] means that if someone instantly measured the number of photons using some sophisticated device, they would find 957 with 60% probability and 959 with 40% probability.

Now let's consider a single rod cell, one that might be hit by one of the photons. It also has a bunch of possible states, but again, for simplicity, let's boil them down to an "activated" state [act] and a "nonactivated" state [nact]. Before the photons hit the eye, the rod state is a superposition of the two, but with the [act] state having very low magnitude, signifying that the rod is basically unactivated. So, the state would look like this:

$$[Rod\ state] = [nact] + {}_{[act]}.$$

Before the photons hit the eye, the rod state and the photon states are pretty much unrelated, and we can think of them separately. Quantum mechanics has a way of describing states that are separated like this: you simply multiply them together. Thus,

$$[Combined\ rod\ and\ glint\ state] = ([957] + [959]) \times ([nact] + {}_{[act]}).$$

Since the state of the glint and the state of the rod each contains two terms, multiplying them gives a combined system that contains all *four* combinations of activated and nonactivated rod, and 959 or 957 photons, which we can write like this (with multiplication signs implicit where two states are next to each other):

$$[Combined\ rod\ and\ glint\ state] = [957][nact] + [959][nact] +$$
$$[957]_{[act]} + [959]_{[act]}.$$

Although the rod and the glint appear to be connected in some way, this is an illusion: the combined rod and glint state is the very same state of two separate systems, since it can be refactored back into the preceding version, and they will remain separate as long as they don't interact.

But, once the photons hit the eye, pass through the cornea, and dodge all the optic nerve connections, we can no longer treat the photons as separate from the rod; there is a relation induced by the interaction between the photons, the various layers they pass through, and the rod cell. This interaction is, by the evolved structure of the rod cell, of a very particular type that *correlates* the rod cell state with the glint state. This means that the interaction changes the state in such a way that there is a much higher magnitude for states that include an *activated* rod than for those that don't, and also such that there is extra magnitude for the terms that include 959 photons and an activated rod, as compared to 957 photons and an activated rod. Thus, after interaction the state might look like this:

[Postinteraction rod and glint state] = [957][nact] + [959][nact] + [957][act] + [959][act].

Roughly speaking, the photons have activated the rod (the *[act]* parts are bigger than before), and 959 photons have activated it more than 957 photons have. Yet, that's only *sort of* the case: we still have a superposition, and the superposition still has four parts! The form of the interactions in quantum mechanics can change the amplitudes accorded to states, but it does not actually change the states and cannot make terms in the superposition disappear entirely. The rod has "seen" the photons in the sense that the photons and rod have interacted, but all four possibilities are still in the system: nothing has "happened" for sure yet.

Now let's consider the rod's interaction with the ganglion nerve cells, which might be in a "firing" state *[firing]* or a "nonfiring" state *[nfiring]*. Here again, interactions provided by the retinal system both couple and correlate the rod cells with the nerve cells, so that the magnitude for *[firing]* is higher than before the interaction with the rod, and the magnitude for terms including *[959]* is higher than for *[957]*. But again, *all the terms in the superposition are still there.*

And so it continues through the ganglion cells, the optic nerve, and the visual cortex: each interacts with the last to create correlations, with the end result being correlations between the state of the visual cortex and 957 versus 959 photon states. If we represent visual cortex states of "saw" *[saw]* and "didn't see" *[nsaw]*, then after four successive stages we would end up with $2^5 = 32$ terms in our superposition; one of these might look, for example, like *[957][act][firing] [nerve][nsaw]*.

This is a big mess of a superposition, but we might boil it down by categorizing these terms according to just the *[saw]* or *[nsaw]* states and the photon number states, reducing it to something that looks like this:

[Postseeing state] = *[957][nsaw]* + *[959][nsaw]* + *[957][saw]* + *[959][saw]*.

Just as with the rod, the relative total size of the *[saw]* and *[nsaw]* components indicates that the packet was just at the threshold of visibility. The facts that *[959][saw]* is larger than *[957][saw]*, and *[957] [nsaw]* is larger than *[959][nsaw]*, express that it is easier to see 959 photons than to see 957 photons.

But which thing actually happens, and at what point? There's no Munenori who both sees and does not see the glint. He sees it or he doesn't. He lives or he doesn't. You noticed that sound just now, or you didn't.

There's a subjective side to this question: *Why is there no Munenori with both experiences at once?* This pertains to the personal experience of what quantum physics says is a superposition of two very different brain states that should correspond to two very *different* personal experiences. We can make a lot of headway by asking more generally: How can the quantum, superposed nature of a system be lost?

We saw when THROUGH THE GATES that the crucial difference between the classical and quantum behaviors of particles is *interference*. We can't just add the probabilities of different possibilities; we have to add the *complex numbers* assigned to the states (which include both magnitude and direction), then find the magnitude of the result. Our equation for the postseeing state described earlier *looks* as if it says: "There's a probability given by the size of *[957][saw]* and a bigger probability given by the size of *[959][saw]*, and so on."

But it doesn't actually quite say this! Once THROUGH THE GATES, we saw that the magnitude of the sum of two complex numbers is not the sum of their magnitudes: there is interference when the spinner arrows do not align. Mathematically, it turns out that those "interference" bits would be represented by expressions involving, for example, both *[959]* and *[957]*, or both *[saw]* and *[nsaw]*. These interference terms express something distinctly quantum, and we can regard a system as "acting quantum" to the degree that these terms are substantial.

Now we come to a beautiful insight formalized by Erich Joos and Dieter Zeh in the mid-1980s. Although we've pretended that Munenori's retina, optical nerves, and so on, are all isolated systems, they of course are not. They are interacting with all kinds of stuff in their environment. In a process now called *decoherence*, these effectively random interactions, when folded into the state of the combined system and environment, cause the interference terms to cancel out to very high accuracy. As Joos and Zeh put it: "The interference terms still exist, but they are not there."[14]

This is an instance of a general truth: once you put a quantum system in contact with an environment that is complicated enough, the system's quantum nature, and in particular *quantum superposition*, becomes impossibly obscured. Once two possibilities have "decohered," any new system interacting with the decohered possibilities for the old system will create a combined decohered combination of the new system with *each* version of the old; the description has thus effectively split into two (or however many) distinct "worlds." Thus the superposition of an activated and an unactivated rod cell would quickly become a decohered one if the rod cell's environment were included; once it interacts with the ganglion cells, those in turn form a *decohered* superposition of ganglion cells that both have and have not received signals from a rod. And so on. In this way, following the formalism of quantum theory results in a superposition of many different brain states. Crucially, though, *none* of these states describe anything like a strange, half-dead, half-living zombie Munenori. Decoherence makes the split a clean cut.

THIS IS SATISFYING, but it raises another somewhat distinct and much thornier question: *How does the combination of four possible Munenori-photon combinations turn into a single Munenori that either lives or dies?* That is, how does a superposition of *potential* outcomes connect with a single *actual* outcome? The quantum state is supposed to be a full, complete description of reality, containing all the information about the system there is to be had. Yet after decoherence into two superposed-but-not-interfering possibilities, it seems there is a major mismatch between the state, which has two possibilities, and "reality," which has just one: Munenori lived *or* didn't. You noticed the sound, *or* didn't. This major mismatch is what most people refer to as the *quantum measurement problem.*

There are approximately as many approaches to the quantum measurement problem as there are people who have seriously thought and written about it; it's an amazingly subtle issue. We can dramatically oversimplify things usefully, however, by dividing most of the approaches[15] into two camps that might be called the *epistemic* and the *ontic.*

The epistemic approach says that the quantum state (or wavefunction) is a mathematical description of *everything an observer can know about a system.* It is akin to the probability P of a roll of a die landing with the 6 side up: different observers such as you or the simulator can meaningfully attribute different probabilities to the roll. But after the roll, everyone agrees that the probability $P = 100\%$ for whatever roll came up, and $P = 0\%$ for the others. In a similar way, the epistemic interpretations would hold, the wavefunction encodes *full knowledge of the system* according to an observer accessing it. Before measurement, the wavefunction attributes some probability to various outcomes. After measurement, all of the probability shifts to the outcome actually observed by the observer. In this view, if we think of the chain running from photons to retina to nerves to brain, leading to a superposition of Munenori's brain configurations, we are free to describe any of them as a superposition; but in the mind of Munenori, only one is actually perceived and provides a basis for action. Replacing a wavefunction with a new one that reflects new information acquired by an observer is often denoted *collapse of the wavefunction*, though this term tends not to be used for probabilities, where we would just say "changed on the basis of additional information." The epistemic view, though, is that these are essentially the same.

The ontic view is rather different in spirit. It holds that the wavefunction *is* reality, or at least is in unbreakable one-to-one correspondence with reality. Thus, when the wavefunction splits into two superposed, decohered worlds, we should take that literally as saying the world has decomposed into two different worlds. No part of the wavefunction ever "goes away." Rather, we can talk about relations between parts of the wavefunction—saying, for example: "The state labeled 'lots of photons' is *correlated* with the state labeled 'sees', while the state labeled 'fewer photons' is *correlated* with 'doesn't see.'" Then a Munenori who sees light can conclude that he did so because there were lots of photons, while a Munenori who missed the glint concludes that there were not enough photons for it to be visible. In this view, if we think of the chain running from photons to retina to nerves to brain, leading to a superposition of Munenori's brain configurations, we should then continue the chain, acknowledging that the brain interacts with the rest of the head, the body, the air around the body, the grassy field in which Munenori stands, the city of Kyoto, and so on. The superposition just grows to encompass more and more.

Thus has the world split into a world that splits into many worlds, and a world that is what is known of the world.

15

WHAT CANNOT BE KNOWN

(ZUIŌ-JI TEMPLE, JAPAN, 1627)

"Do you think," you asked Umpo Zenjo as you sat in the garden, "the Buddha really did know everything, as in the sutras?"

"Oh," Zenjo replied, "the sutras are very wise, but they are also quite old stories not to be taken so very literally. I think the Buddha knew enough, do you not?"

"Yes, but do you think it's *possible* to know everything? For an enlightened being? Or a God?" (Or a djinn, you think to yourself.) "How would you find out, and measure it all? Where would you store the information? What . . ."

"I think you know *too* much!" Zenjo laughed. "Shall I pour you a cup of tea?"

Today you find a slim, dusty volume on your table. It's entitled *What Cannot Be Known*, and you half expect it to be blank pages.

Instead, you find a book of poems.

It is sometimes said that nothing is impossible. In a sense, that's true. Even tasks that appear extremely difficult can often be accomplished by a combination of ingenuity, luck, and hard work. The complex interplay of events and causes in the world certainly generates a lot of unexpected surprises, and often the rules that might *forbid* something from happening are unclear, or brook exceptions. Quantum mechanics provides many such exceptions, particularly on very small scales: behaviors that are impossible according to classical physics often *are possible*, albeit with some very low probability, when quantum effects are considered.

That being said, there are things that, according to our current (and, it should be said, incredibly effective) laws of physics, *cannot be done* or, in the phrasing of Zenjo's old book, *cannot be known*, as told in three pages of poetry.

PAGE 1.

On a knife's edge, the monk waits for the ringing of a distant temple bell.
A dove passes overhead.
The moment, decided, dies.
The bell sounds, too late.

The first impossibility is that *signals cannot be transmitted faster than the speed of light*, or, put differently, we cannot know about anything that occurred outside of the space-time region called our *past light cone*, from which signals could reach us traveling at light-speed or below. A monk at a given place and time—say, just as he makes a fateful decision—may or may not be able to hear a distant temple bell ring; the sound may arrive too late. Given a telescope, he may *see* the bell ring slightly before he hears it, because light travels faster than sound, and thus *know* that the bell has rung even if he has not yet heard it. But if light itself cannot reach his given place and time quick enough to inform him, then, according to Einstein's relativity, he *cannot know* that the bell has rung at that event.

Why is obtaining this knowledge impossible? There are several arguments within special relativity to this effect. Probably the most convincing is that it can be shown that if two agents, A and B, in different reference frames can both send signals faster than light, then agent A can send a signal to B and have it sent back so that it arrives before A sent it! This is a paradox of a most unpalatable sort—almost exactly the type that arises if you imagine traveling back in time and destroying the plans for your time machine. Such paradoxes can make great drama, but little sense. This provides a strong case that if special relativity is correct, then either signals must be confined to sub-light-speed, or else something is terribly, terribly wrong with the way we imagine doing things in the world (such as deciding whether to send a signal)—to such a degree that it would honestly be hard to make sense of anything.

What's particularly interesting here is that this limit applies to *any* sort of signal, using any types of particles, fields, telepathy—what have you—as long as the rules of special relativity are in effect. The limit appears deeply built into the structure of reality that knowledge is *local*, and that there is a large part of the Universe that you simply cannot directly observe.

PAGE 2.
Day and night the forger labors.
Failed brushstroke after failed brushstroke,
the copy does not live.
Worse, the stain of failure spreads,
and all is lost.

Although quantum mechanics appears somewhat lax in terms of what it allows, we have already seen that it can be quite strict in terms of what we can *know*. We can't know the position *and* velocity of a particle precisely at the same time, or which way it is facing *and* how much it is spinning, and so on. These impossibilities rest on the idea that a particle is *too simple* to give definite answers to more than a couple of questions. If we ask just the right questions, we get definite answers; but if we ask the wrong questions, or too many, we get indeterminacy and uncertainty.

Now, when you contemplate this a bit, it would seem to provide a way to circumvent quantum uncertainty: *just ask the right questions!*

That's a great idea, if only you could know what the right questions are. But knowing the right questions means that you know quite a bit about the quantum state. And to know about a quantum state that you yourself did not create means you have to ask questions. But you don't know the right ones. So . . .

What would be helpful would be a procedure whereby you could *copy* the quantum state of an object without asking any questions. With a device carrying out this procedure, you could make many copies and ask questions of those copies without disturbing the original; thus, you could *know* its quantum state without asking any questions that mess that state up.

Alas, such a device turns out to be impossible. A "no cloning" theorem to this effect (and not much more complicated in its detailed statement) was proved in the mid-1980s and is a core result in quantum theory. It states that you cannot have a device—that is, any valid procedure for evolving a quantum state—that can turn a quantum state ψ into a pair of states $\psi \times \psi$ representing two copies of the same system. Quantum states are unforgeable works of natural art, and if you try to copy one, you will either fail or (perhaps hopelessly) mar the original.

The fact that you cannot make copies of a quantum state means that if you come upon an unknown quantum state, you *cannot* determine what it is. Of course, you can obtain partial information about the system by making measurements, but you then unavoidably change the system. The only way to measure *without* effecting such changes would be to know the right questions to ask, but this would require knowing the state itself. And it cannot be known.

PAGE 3.

In a dewdrop the Sun hangs from a thread, and shakes as
a spider struggles, ensnared in another's web.
Even the greatest skill cannot untangle their strands from all others.
And a blade can but cut the strings.

Even if we cannot know the quantum state of a system that we come upon in the world, it's nice to think that the system at least *has* a quantum state—that there *is* some way that the system "is." Except that in general, there isn't.

While SPLITTING THE WORLD, we saw that via its interactions, the quantum state of a gleam of photons became *entangled* with the state of cells in Munenori's eye. This entanglement means that we can no longer accurately describe the photons as having their own state: measurements or other effects on the cells also necessarily affect the photons, because they are bundled into the same terms in the quantum superposition.

At some level, entanglement happens whenever one system interacts with another one. And the forms of Newton's, Einstein's, and Maxwell's equations mean that pretty much everything is interacting all the time with pretty much everything else. That immediately raises two questions. First, how is anything *ever* unentangled? Why talk about *the* state of a system at all, if you can never actually say what it is? Second, how can we predict what a system will do, if we can't assign a state to the system? Do we have to predict the whole universe to see what will happen to a grain of sand?

For the first question, we can observe that any interaction that happens can, *in principle*, be undone. That is, if there is a way to entangle Munenori's retina with photons, then we should be able to reverse this interaction so as to "unentangle" an entangled retina-photon system into two separate ones. This unentangling process requires carefully designed procedures; in a laboratory, we might, for example, create a single isolated quantum system, like an electron spinning around a vertical axis, isolated from its environment. This careful procedure has evolved the electron-world system into an electron-world pair, so that we can, at least for a time, describe the electron as its own system. The bad news is that this isolation is very hard to achieve, and very, very hard to maintain. Indeed, this is the core challenge in building *quantum computers*, which rely on maintaining an isolated quantum system that does not become entangled with its environment. Because even the tiniest interactions create entanglement (and, almost unavoidably, decoherence), this is tremendously difficult in practice. Isolating a system is like escaping from a spider's web; keeping it isolated is like dancing in the midst of one. You have to be very careful indeed.

That brings us to the second question: If it is so very difficult to disentangle one system from another, how can we ever use quantum mechanics to predict anything? Well, when a tangle is too difficult, you cut the string! There is a procedure we can apply to (mathematically) *excise* a section of the world to form a system of our choice, acknowledging that because our system is entangled with others, and because we have chosen not to keep track of those others, there will be added uncertainty in the description of our chosen system. That is, after applying this procedure, we can say only that the quantum system is in a given state *with some probability*. We might, for example, end up with a situation in which there is a 50% probability of being in state [→] and a 50% probability of being in state [←]. This is *not* to say that the system is in a superposition [→] + [←]; that would still be a single quantum state that constitutes a definite answer to some question. In contrast, our 50-50 mixture of states, while it does give probabilities for observations, does not provide a definite answer to *any* question.

The cost of isolation is ignorance.

16

WHAT WE TALK ABOUT WHEN WE TALK ABOUT FREE WILL

(ARABIAN DESERT, 1610)

By the 12th day in the caravan you are just about ready to head back to the djinn's cave. At least it was cool there.

The only compensation for the endless searing heat and camel stink is the conversation; as luck would have it, you have fallen in with some rather academic students, who delight not only in helping you learn Farsi, but in debating the nature of free will—a topic of recent and intense interest for you.

After deciding against discussing the djinn directly, you start by going straight to the heart of the matter, in terms of religion: If your decision is already known to God, and thus determined, how can you say it is your decision? You put this to the students:

"When I decided, I feel that my decision making. But if my decision is already God known and thus, determining how the decision I can say?"

You're just learning Farsi, remember?

One of them looks a bit puzzled but attempts a reply: "If the reason that you confirm you can feel their decision is free. Even if the set outcome fixed."

You're not satisfied by this, and try again: "But suppose that God or any decision of the agent you are using the full power of anticipating wisely you can. Then all internal meditation importances not."

Your companions talk among themselves too fast for you to

follow, then one replies: "Maybe not the same free will in order to: a definition to be the absence of external coercion. The second definition based on your decisions made for reasons which you will understand. Perhaps the third definition that your decisions other than the fact the decision can't. The final definition of disability for their decisions to be predicted, in principle."

And so it goes, through countless hours.

> This causes a very absurd thing. If someone said that without the bones, nerves and all that you will not be able to do what I've decided, modify, but definitely not to say that the cause of what and who I have chosen the best method, even if you are in my mind, to talk to lazily and carelessly. Imagine being able to detect the real cause without which the cause as the cause will not be able.
>
> —Socrates, in objection to materialist explanations
> of his choices (poorly translated among
> English, Greek, and Persian)

Discussions of free will often seem to go like this: what is so clearly true to one is so clearly false to another, as if the two are speaking wholly different languages; and everyone is confused. We may do no better here, but perhaps by teasing apart some of the issues, we can say something useful about some of them.

Let's start with the *experience* of free will. It has multiple possible elements. In feeling free, we feel that we are not coerced or controlled in making a particular choice. Prior to the choice, we feel that we have multiple real options, any of which we might choose; after the choice, we feel that we could have chosen differently. We feel that we make choices partly by instinct or intuition, but at least partly for *reasons* that make sense to us, that we identify with, and that we provide when asked *why* we made the decision we did. We feel that the decision process is one that we must work through; almost by definition, we can't know what we will decide until we have actually decided.

If any of these were violated, we would not feel free: when coerced, we would feel unfree; if we saw only one choice, we would feel unfree; if our choices happened at random, regardless of our wants or hopes, we would feel unfree; if we knew our decision immediately, we would not even feel that we were deciding—we would have decided!

Most of the time, we experience none of these unfreedoms. We agonize, or don't, over decisions of true import. We bear the weight of responsibility, possibility, and regret. This sense of ultimate inner freedom can feel both terrifying and elating in its power. As Sartre put it: "Once freedom lights its beacon in man's heart, the gods are powerless against him."[16]

Yet the djinn tells us it is an illusion. "You are made of atoms," it says, "and I know the worldline of each one, tracing its extreme action through space-time." The djinn knows your initial conditions and evolution equation, and thus the state of the sound-filled air leaving your mouth 12 seconds hence. It claims to know your decisions. Perhaps it can even change them via the flick of the right atom.

How would you feel if confronted by a djinn who spoke your words instants before you? Or who forced you to make a series of gut-wrenching choices, then showed you a parchment on which each one had been written before you made it? What if you knew such a parchment existed, even if you could not ever view it? Would you feel free? Would you *be* free?

THESE NOTIONS of the freedom that we may or may not have are captured succinctly in the four definitions provided by the Sufi student:

"One," says the (properly translated) Sufi, "is the absence of external coercion."

"Second," he continues, "is that your decisions are made for reasons that you endorse."

"Third," he goes on, "is that your decision could have been otherwise."

"Fourth," he concludes, "holds that neither you nor anyone else can reliably predict what the decision will be."

The djinn does not claim that your decisions will feel coerced, and it is perfectly comfortable with your feeling that the decisions are based on good reasons. The djinn simply claims that you *are* coerced by Maxwell's equations and the space-time metric into one inevitable decision, and that your so-called reasons are confabulations—like the patient of Freud, hypnotized to pick up an umbrella, who later does so for the perfectly good reason that it just might rain, even if the sky is clear.

But is the djinn's claim true? We have reason to be skeptical. We know there are things that we cannot predict and that, as we learned from Zenjo's book, cannot be known. To be very precise, let's tackle the Sufi's fourth criterion, and the central claim of the djinn: that by knowing the state of the world in exquisite precision, it can predict everything that will happen, including events in your mind and actions you choose to take. Addressing this question ties together many of the concepts we've been assembling.

To start, consider first a diagram encoding the djinn's claim (see the figure on page 128). In it, there is a world "now," represented by a quantum or classical state spread across space at an initial time t_i. This world is known in full detail to the djinn, as are the mathematical laws of physics. Given both ingredients, the djinn also knows the "future" world, a different quantum or classical state at the later time t_f. This later state includes your decisions, "chosen" actions, and everything else. This setup is generally what is meant by physicists

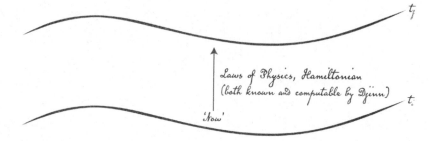

A deterministic world.

and others when they say that "the world is deterministic," and it is what the Pen would have been writing when it is said that from it "flowed everything that would happen until the Day of Judgement." Yet it is far from the full story.

While keeping physics classical rather than quantum, let's now admit that the djinn is a physical system situated within the world, and incorporate relativity, which dictates that some regions of space-time can affect only some other regions, as required by light's finite speed. Relativity adds some crucial elements. In the next figure (opposite), the djinn has an overall worldline, which, upon examination, would be composed of a tangled braid of individual particle worldlines weaving together with others traveling to and from the environment. Your bundle of worldlines would be similar, if rather simpler than the majestic djinn's. Consider now you, the djinn, and the rest of the world "now." As in the previous figure (above), and as when we were first RELEASING THE DJINN, this is all of space at some rather arbitrarily labeled set of space-time events that we could call "now." The later time t_f is a similar, later region of space-time.

Now we can ask: "What does the djinn *now* know *for sure* about the 'world *now*'?" If by "know for sure" we mean actually observe, then the answer is: nothing whatsoever. The djinn observes *earlier* times, via information transmitted to it at the speed of light or below. It can, in principle, know *nothing* about what is happening at the same time but at some distance.

Consider, then, some earlier time t_i. At this time there is some

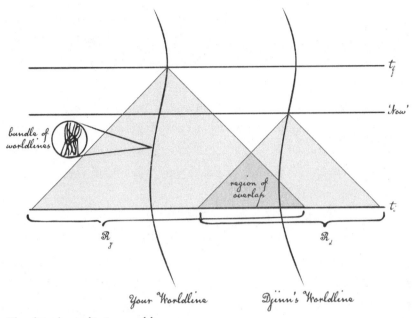

bundle of
worldlines

region of
overlap

t_f

'Now'

t_i

R_y

R_d

Your Worldline   Djinn's Worldline

The djinn's prediction problem.

finite region of the world, which we may call R_d (on the right in
the second figure, above), that the djinn *can* know about; it is tran-
scribed by the djinn's past light cone. Let's assume that by dint of its
extraordinary abilities, the djinn can gather unlimited and infinitely
precise information about all of the matter in this region: a perfect
specification, say, of the type, position, and momentum of every particle in it.

But consider now your actions at time t_f. What is necessary in
order to predict them? What would suffice would be perfect knowledge of everything at the earlier time t_i that lies within the past light
cone of your location at time t_f; we'll denote this region R_y (on the
left in the second figure). Why this region? Well, particles outside of
this region can't matter: if they did, their influence would be faster
than light. But everything within this region does, in principle,
matter, and can *potentially* be important. Any point in R_y could contain a particle that might, via its interaction with you, help change
your mind one way or the other.

We now see an immediate problem for the djinn: the region R_y (over which it needs to know everything) only *partially* overlaps (in the shaded region of the figure) the region R_d (over which it *can* possibly know anything for sure). So, strictly speaking, the djinn simply cannot predict what will happen at your position at time t_f. In fact, strictly speaking, the djinn cannot predict *anything at all*; every event to the future of the djinn "now" has a past light cone that encompasses regions outside of the region over which the djinn can gather information "now."

So, are we done?

No, we're just getting started! Now that we've rolled dice, split the world, and read bad poetry, the idea of the djinn gathering unlimited precise information should be ringing mental alarm bells, because quantum physics, sidelined so far, places rather strict limits on that information gathering. Let's look at those limits carefully.

Consider the region R_d that is accessible to the djinn "now." Suppose that the djinn knew the precise quantum state of this region at t_i. We saw from SUFFICIENT REASON FOR A ROLL OF THE DICE that this state contains sure answers to a very particular set of questions. But there's no guarantee whatsoever that any of these correspond, either at t_i or after evolving the state to any later time, to questions like "what will this miserable human decide?" Rather, there is every reason to think that such answers are *not* in the list, as we saw when rolling the dice: prediction of what is going to happen generally requires knowledge of multiple quantities, such as position and velocity, that cannot be simultaneously known precisely. It is far more likely that through time, evolution, and interaction with the environment, the state of region R_d evolves into a superposition of macroscopically distinct outcomes, just like the outcomes of Munenori's sword fight. So, for questions like "what will be decided?" even a precisely known initial quantum state is very likely to evolve into *multiple* distinct possibilities. One of these might be much more probable than the others, or not.

This situation is as if the Pen were to write page one of a story, but then on page two, write two different continuations of the story, where the even-numbered words on the page would be one continuation, and the odd-numbered words the other. The third page could continue each of these two different ways, using every fourth word for each version, and so on. The result would be a single book, but containing many stories, bifurcating the plot on every page.

Moreover, all of this assumes that the djinn knows the state of R_d, and there are two major obstacles to this. First, the no-cloning theorem tells us that the djinn can never directly measure the quantum state; that would be cloning it! At best, the djinn can ask some questions of the state and receive some answers, while in the process almost inevitably altering the state itself. Second, quantum theory tells us that the matter in region R_d is almost certainly quite *entangled* with everything outside of it. Thus, we must "cut the web" as enjoined on the third page of WHAT CANNOT BE KNOWN. The result, in general, is a *mixture* of quantum states in R_d: probabilities that we attribute to the different possible states that could describe it. The djinn is thus in a pickle: simply by—and of necessity—picking out an initial region, the djinn can, at best, consider a mixture of states. The djinn then cannot know what these states are and, in gathering information about them, changes them. And finally, as we saw in the classical case, R_d is not even the region about which the djinn needs information in order to predict you at t_f. Rather, the djinn needs the state on R_y, which only partially overlaps with R_d.

To sum up from a slightly different angle:

1. To predict you at t_f, the djinn would need the quantum state on R_y and have the ability to evolve it to t_f.

2. Even given the state at t_f, the djinn would have only uncertain predictions of what you will do, because even a precisely known state evolves into multiple possible sets of answers to questions the djinn would pose.

3. Even if the djinn could access region R_y in its entirety, it could not know the precise quantum state, because of the no-cloning theorem.

4. Region R_y does not even have a single quantum state, because it has been cut out of an entangled state describing both region R_y and its environment.

5. The djinn has only partial access to R_y, being limited to observing R_d.

This seems quite hopeless! The djinn appears quadruply unable to predict you—or anything else—at t_f, at least not in the sort of completely precise, absolutely certain way that the djinn claims.

Now are we done? Not quite. While, *in principle*, nobody can predict anything, in practice of course we can! As the djinn suggests, if you let go of your lamp, it "will" fall. The quotes indicate that this is not absolutely certain: it is *possible* that a blast of laser radiation from ultra far away could intercept the lamp on the way down, and there is no way to have seen it coming. We generally assume that such effects are absent, and we are generally safe and correct in doing so; the world appears to be constituted so that, in large part, simple local cause and effect are not thwarted by new things entering the light cone between when we predict something to happen and when it actually happens (or doesn't). And although quantum uncertainty of multiple types is quite inevitable, there are many systems, from planets to dust motes, in which this uncertainty can be completely ignored with nearly no loss in predictive power over the questions you'd like answers to. That is, although the Pen may write many stories, from a wide perspective some of these stories are much, much,

much more probable than others—enough that we can see almost exactly where the plot is going.

The answer to the question of whether a djinn could, in principle, predict *everything about us* is no. Yet we can still ask: "Is the type information that the djinn *could*, in principle, gather sufficient to predict your internal thoughts and decisions, in general and with very high confidence?" After all, it would still violate our experience of free will if the djinn could do this "just" 99.99% of the time, or only under relatively controlled conditions.

We need, then, to ask: "How much do we really need to know about a region, or a body, or a brain, to predict or understand what a mind will do, or decide?"

17

THE MIND OF MING

Although you have fallen from favor with the khan, you've come upon an unexpected and informative ally in Li Yung-Fang, who has defected from the Ming dynasty. His counsel on the working of the Ming dynasty's government is invaluable in understanding how the khan may defeat it, and you are also taking his lessons to heart regarding your own predicament.

The more you hear, though, the more you keep coming back to the same basic question: How does anything get done, and any decision get made?

As Yung-Fang explains it, it seems simple on the surface: the emperor decides. But the emperor almost always does as his ministers advise. And the ministers act on the basis of reports from the deputies, which are prepared very carefully in accordance with strict rules. The deputies follow the Code of Conduct to the letter (as recorded by the scribes) and receive information from the scholars, governors, generals, and collectors. Each is governed by an incredibly intricate set of rules. And on top of this are myriad *social* requirements ensuring that the place and status of each government member are precisely respected at the right level.

"But this is astonishing!" you say to Yung-Fang over stiff drinks. "The emperor doesn't really make decisions?!"

"Well, he thinks he does, but the truth is that most of them are made well before they reach his chair."

"But made by whom and how?" you ask. "Everyone is just

following a complicated set of rules and pushing around papers! How do the decisions make any sense?"

"Explains a lot, huh?" Yung-Fang chuckles. You both drink to that.

"But seriously," you push on, "the decisions aren't random, and they are sometimes even wise. How can that be so, if they are just the outcome of a bunch of rules being followed?"

"Indeed," agrees Yung-Fang. "The empire is wise, even if the emperor is not."

> It is possible to invent a single machine which can
> be used to compute any computable sequence.
>
> —Alan Turing[17]

Recapitulating our trip with Munenori, if we were to follow the quantum state of the trillions of photons traveling from the well-lit word "recapitulating" into your retina, up your optic nerve, and into your visual cortex, we would encounter a system of inscrutable and almost unbelievable complexity.

Grouping the components of the system's quantum state into things like molecules arranged in complex biological cells, we would see on the order of 20 billion neurons in the cerebral cortex, connected together by approximately 100 trillion synapses, interacting in a dizzying orchestra. Each neuron receives about a thousand synaptic pulses per second and—on the basis of these pulses, potentially other chemical information, and its own internal dynamics— "decides" whether and when to fire. These firings trigger other neurons, and also form a host of coherent frequencies over small and large areas of the brain, which in turn affect the firing rate. A vast array of chemical and even genetic signaling further modifies neural and synaptic operations.

At exactly the same time and in parallel, we can say something like: "Patterns of light and dark have their edges identified within the retina, and are composed into shapes within the visual cortex." These shapes are pattern-matched to "letters" that are further matched to the word "recapitulating." In turn, this word immediately calls up a vast array of associations, meanings, contexts, and predictions. It almost instantly interacts with other words, like "our" and "trip" and "Munenori," creating an experience of a dimly imagined samurai, discussion of photons, quantum concepts, and paradoxes. This quasi-conscious experience floats, waiting for the phrases "quantum state" and "trillions of photons," which are satisfyingly worked into the narrative and strengthen it. The phrase "into your retina" connects you to other sensations of reading, perhaps bringing flickering awareness to your eyes. Then there is a quick mental loop as you notice that "recapitulating" matches the earlier phrase, creating momentary unexpected confusion and interest. Setting this aside, you reach "unbelievable complexity" and probably start to envision a set of neurons, not to be disappointed by the following paragraph.

How are these two marvelous and intricate processes—one of neurons and synapses, and one of *mental processes*—connected? Are they one and the same? Does one "cause" the other? How do synapse firings and dendrite wirings relate to an experience of the glint on a sword, or the experience of thinking about the experience of the glint on a sword that you just had?

This problem is very hard. It's literally called the "hard problem." As David Chalmers, who coined that term, put it:

> Consciousness is the biggest mystery. It may be the largest outstanding obstacle in our quest for a scientific understanding of the universe. . . . It still seems utterly mysterious that the causation of behavior should be accompanied by a subjec-

tive inner life. . . . We do not just lack a detailed theory; we are entirely in the dark about how consciousness fits into the natural order.[18]

Let's start here by asking something that is still fairly perplexing, but somewhat easier: How does a physical system that follows inexorable laws ever *compute* or *reason*? If we ask a computer what 23 + 17 is, the reason it displays "40" is that, per Maxwell's and Schrödinger's equations, the electrons, semiconductor elements, key wirings, and so on, all follow their inevitable course, thus leaving a glowing "4" and "0" on a display. But of course, the computer also displays "40" for the reason that 23 + 17 is equal to 40! How are these two things the same?! And what is the "real" reason? Does the Ming emperor make decisions because they are just and right and wise, or because a hierarchy of brush-pushing bureaucrats carefully follow rules that lead to those decisions? Or both?

This question is still too hard. Let's make it even easier. What does it mean to "compute" something at all? Most generally, we might define a computation as a set of rules that map input into output. Logical gates such as AND, OR, and NOT are perhaps the simplest examples, with both input and output being binary values. Such elements can then be arranged in complicated systems that map very complex sets of binary data to others.

In this sense a physical computer is much like a physical machine: a stable set of material that, following physical laws, reliably produces the same output for a given input. But in another sense it's quite different, because the physical device is just there as a means to *perform* the computation. An AND gate could be created with electrons in semiconductors, golf balls and Tinkertoys, pilgrims on a plateau, organic molecules, or what have you. Just as the integers can count up any type of object, computations can run—and run in just the same way—on any substrate. And just as in mathematics, this means that we can reason at an *abstract level* about computations and what they will, can, and cannot do.

Many general results have been proved about what any possible combination of AND, OR, and NOT gates can accomplish. But it's not the only, or even most widely used, framework for thinking about computation. In the early twentieth century, Alan Turing, Alonso Church, and others created a sophisticated theory of computation based on Turing's model of what is now called a *Turing machine*. Such a machine requires a "tape," which is a reliable readable and writable information storage mechanism, and a "head" that can read and write from the tape according to some finite system of rules. Many things have been proved about Turing machines, among them the somewhat surprising fact that just about *any* computation you can imagine can be done on a suitably designed and programmed Turing machine, as simple as the device seems. It might be extremely inefficient (you'll need a lot of tape!), but it is, in principle, possible.

For example, the action of any set of logic gates on a string of input bits can be computed by a suitable Turing machine. Because this is the basis upon which microprocessors and other modern computer elements operate, it immediately follows that anything that any standard digital computer does is equivalent to some Turing machine.

Another computational framework is that of quantum computers, which exploit the complex amplitudes for quantum states to perform certain types of computations much more efficiently than classical computers do. But although more efficient, quantum computers can still be *modeled* (i.e., substituted by) a much less efficient classical computer, and hence by a Turing machine.

THAT THE PHYSICAL LAWS of our universe allow computers is not a given. The stability and regularity of fairly simple physical laws enable a given computer to produce the same output time after time.

Likewise, nature allows for physical systems that are sufficiently self-contained to follow *internal* dynamics, generally free from outside interference and connecting only via "input" and "output" to the external world. The universe didn't have to be this way, but it is! This stability and independence create, in a real sense, a new level of reality that is defined, understood, and constrained by the rules of logic and computer science rather than the rules of fundamental physics, which it *also* obeys. For a prediction of what a Turing machine made of Lego bricks will do, each step of its operation *could* be understood using the laws governing the bricks. But why bother with that approach, instead of just tracking what the computation itself should do? And what if someone told you what rules a computer was applying to its input and output, but refused to tell you how the computer was physically instantiated? You would be fine, because you *don't need to know.*

So, if what is going on in a complicated physical system is best described as a *computation*, then perhaps it is far, far easier to predict than it might seem. We don't have to understand every atom—just which computation the system is actually doing.

Is your mind a computing device like this? It is clear that the human mind *can* do computation: we can perform all the operations of a Turing machine, or decide the output of a set of AND/NOT/OR logic gates. We're not terribly good at this, though, with any human outclassed by a dirt-cheap solar-powered calculator; we have to work hard to force our minds to step through well-defined algorithms. In their more typical operations, our minds work more like a large and efficient bureaucracy, or perhaps a modern sophisticated software system. There are individual units that are functionally and sometimes physically identifiable, for processing and integrating sensory data, taking motor action, controlling internal bodily processes, regulating hormonal and other chemistry, creating and recovering memories, visualizing and predicting future actions, and so on.

We tend to feel that we are at the top of a hierarchy, like the Ming emperor, and in a sense that's true: the decision to do something, experienced as "you decide," takes charge of the whole operation in a fairly clear way. If you decide to close this book, a single imperial command sets in motion a whole variety of mental and physical processes—almost all of them too complicated to follow in detail, were you actually to try—that lead to the book being closed. For the most part, the system works beautifully and you're generally aware of it only when, for one reason or another, your edict fails to create the desired outcome.

At the same time, you can ask: "Where did that decision to close the book come from?" There could be many causes, most probably unknown to your conscious mind: there's fatigue; there's a mounting level of distraction in the room; there's the increasing urgency of some responsibilities; there's hunger or thirst; there's a need to satisfy some low-level addiction to email or media; there are some minor physical discomfort nags. All of these combine with complex weightings that, at some point, accrue to the side of "time to close this book for a while." This, then, pops into your conscious mind, and you probably then write a story as to why you made that decision, like "I should make some coffee," that might partially cover the list of actual causes and reasons, and might have almost nothing to do with them. So, then, was it the emperor who really decided, or the bureaucracy?

Also like an imperial bureaucracy, mental output is variable and somewhat unpredictable. A peon in the Ming government may feel quite powerless to change government policy, but if some policy decision is difficult, its resolution may hinge on the reports from lower levels in the bureaucracy. Those reports contain decisions that may be based on reports from even lower levels, and so on; thus, changes in a peon's work might very much influence a high-level decision. Our minds appear quite like this also, with a significant stochastic component entering and affecting our thoughts.[19] High-level decisions are stable in many senses: almost no functional, sane person will "randomly" decide to step into oncoming traffic or go on a murderous rampage. But *hard* decisions might depend on rather fine-grained, low-level mental details. And streams of small decisions—like those we make when assembling a sentence—are hard to reproduce even if we try to. We can (and if asked, will almost unavoidably) compose a number of different sentences expressing essentially the same meaning but differing in syntactic detail. These variations are "random" in that we cannot really predict or reproduce them, and they probably depend on variable and stochastic mental processes filtering from lower levels of the mental hierarchy. But they are also not *really* random; the sentences still make sense!

Such, probably, is a lot of our thinking.

HOWEVER DIFFERENT the functioning of the human mind is from a Turing machine, though, we might still ask whether it secretly *is* one. The operating system of a tablet computer, for example, does many complicated things, has submodules for storing and retrieving data, getting sensory input, creating output, and so on. And it does not obviously function like a general Turing machine: would you have an easy time getting a tablet to execute a simple algorithm of your devising? Yet underneath, we know that it certainly *is* running an algorithm (we can look at the source code!) and can be emulated, in principle, by a Turing machine. It also has fairly high-level rules that let you predict with excellent accuracy what many "sensory" inputs—like touching its screen—will do in a given situation, without mucking about with electrons and Maxwell's equations.

Is the human mind also like this? If it is, then perhaps the djinn is not so crazy to think it can predict everything you will do. It just has to quickly predict the output of whatever computation your neural computer is performing, given the input, before you actually perform it. That seems much easier, or at least possible, right?

18

A HALTING PROBLEM

(DAMNED ARABIAN CAVE, 1610)

"Are you ever going to let me go?" you plead with the djinn.

"Well," says the djinn smugly, "that depends on what you do. I shall decide whether you deserve freedom."

You don't want to anger the djinn, but you're very frustrated: "But you already know what I am going to do, right?"

"Well, you haven't done what you're going to do yet, so I have not decided," replies the djinn.

"But if you know the future, you know what I'll do and what you'll decide," you object. "And even if you won't tell me, can't you at least tell me when you will make your decision?"

The djinn considers this for a while, then for a while more.

You see the djinn frown in concentration, then a series of expressions begins to run across its face.

After a minute or two, you start to wonder whether the djinn is ever going to answer, or decide.

After another minute or two more, you realize the djinn is not really paying any attention to you, and you quietly back out of the cave.

Any decider whose decision making process can be described using a systematic set of rules (e.g., the laws of physics) can not know in general beforehand whether she will make a decision and if so what it

> will be. . . . It is less efficient to simulate yourself
> than it is simply to be yourself.
>
> —Seth Lloyd

Sixty-three billion seconds prior to meeting a wanderer in its cave, the djinn discovered something rather vexing. The djinn has always enjoyed running algorithms inside itself, spawning a small dollop of Djinnium to perform a sequence of instructions on a given supplied input *I*, and receiving from the blob of Djinnium an output *O*. Over the course of many such experiments, it found that the blob would generally either spit out the result quite quickly (Djinnium is very efficient), or just sit there basically forever, if the algorithm had some sort of infinite loop in it. Usually the djinn could identify such a loop right away, but with the more fun and interesting algorithms, it could be surprisingly difficult to determine whether the Djinnium would just sit there or not. Moreover, it was not always a black-and-white question; occasionally, the Djinnium blob would sit there for an incredibly long time—as much as full minutes—then suddenly spit out an answer.

So the djinn decided to solve the issue once and for all by carefully crafting a program it would call *H* that would run in a dedicated chunk of Djinnium. This *H* program would tell the djinn whether some other algorithm would behave well or not. *H* would spit out the result "SATISFACTORY," if the algorithm in question, when run on input *I*, would provide an answer (eventually), or "BORING" if it would just sit there forever. That is, *H(A,I)* would be an algorithm that takes both *another algorithm A* and also some input *I* together as input, and spits out "BORING" or "SATISFACTORY" as output. The djinn was quite pleased with this idea, and commenced development and testing of better and better *H* algorithms.

Constructing *H* proved to be more difficult than the djinn had hoped, but it couldn't understand why. So it started thinking in more general terms, *assuming* that it had an *H* algorithm, and then analyzing how it would behave in various situations. This was an enjoyable exercise that occupied many seconds. During Second 43,123, the djinn discovered the fun twist of applying *H* to an algorithm *A* and *also* using that algorithm as the input to *H*—in other words, evaluating *H(A,A)*. Among the djinn's many experiments with this, during Second 43,645 it made up a funny little algorithm *M(I)* that included this trick:

Step 0: **Accept input I.**

Step 1: **Call *H(I,I)*.**

Step 2: **If "SATISFACTORY" is returned, then loop forever.**

Step 3: **Otherwise, return output "THE DJINN IS THE BEST."**

The djinn enjoyed the perverse quality of this algorithm, which does something boring if the input algorithm *I* is satisfactory when run with itself as input (i.e., if *H(I,I)* returns "SATISFACTORY"), and something quite satisfying if the input *I* gives the output "BORING" when run on itself. This was fun to play with for about 23.4 seconds. Then, at Second 43,669, the djinn made a fateful step: it considered what would happen if it fed *M* itself to *M*.

The djinn reasoned that anytime *M* is run on some input, there are two possible outcomes: loop forever, or output "THE DJINN IS THE BEST." So it considered them in turn.

First it supposed that *M(M)* returns "THE DJINN IS THE BEST." That would be quite satisfactory. In fact, it's genuinely satisfactory: since *M* didn't run forever, that means *H(M,M)*, which specifically tests whether *M(M)* finishes, would return "SATISFACTORY"! Buuuut . . . in program *M* itself, if *H(M,M)* were to return "SATISFACTORY," then *M* would loop forever. But . . . *M* didn't loop forever: by assumption, it output "THE DJINN IS THE BEST." Uh-oh.

So, reasoned the djinn, M must loop forever when given M as input. But *that*, it realized, would mean that $H(M,M)$ would return "BORING," and thus M would output "THE DJINN IS THE BEST." Uh-oh.

The djinn hated, hated, hated, *hated* paradoxes. But it could not see any way around this one. As soon as the djinn assumed it could write an unerring algorithm $H(A,I)$, the immediate implication was that it could write an M algorithm with paradoxical output when applied to M itself. The djinn's only recourse was to admit to itself that writing a perfect $H(A,I)$ algorithm is impossible.

The djinn raged against this conclusion; it was bad enough being stuck in a tiny lamp in a tiny cave, without being told by brute logic what it could not do. But over the ensuing 63 billion seconds, it came to accept that if it wanted to figure out whether an algorithm was satisfactory or boring, it might as well just hand it off to a blob of Djinnium, wait, and see. (About 73.3 billion seconds later, the human Alan Turing would come to a similar conclusion.)[20]

WHEN A TIRED and thirsty traveler entered the cave and asked the djinn what the djinn would decide, or at least *when* it would decide, the djinn was so intent on using the human for its own purposes that it didn't notice itself falling into its own snare, built of self-reference.

When the human said: "You must know what I'll do!" the djinn agreed, so it whipped up a relatively simple trial algorithm D that, given some input I, would output the decision to which the human would come, given the situation described by the data I. The djinn started the routine running in a chunk of Djinnium, but a result was very slow in coming, and the djinn realized that the algorithm might be stuck in a loop; but it was not obvious why. What a waste of time! So the djinn's next instinct was to develop another algorithm, H, to fix this: it would analyze D and I, return "DON'T KNOW" if $D(I)$ would fail to finish, and otherwise return simply $D(I)$.

The djinn frowned. Its excellent memory had just dredged up the recollection that it had been down this road before, 63 billion seconds earlier. The function H that the djinn had just planned to invent was pretty much exactly the same function H that it had tried to create all those seconds ago; it even used the same letter! Thus, creating this function was in the hated set of things the djinn could not do.

The djinn became even more irritated as it realized and proved several implications and extensions of this result. The most obvious was that there was no simple and safe shortcut to predicting the human's decisions, *even given* perfect understanding of the algorithm that the human was presumably running in its tiny brain in order to decide; the djinn would have to run the algorithm itself and just hope that it finished at some point. Of course, figuring out the exact algorithm might be challenging, so the djinn had planned on using something much simpler; surely it could do the human's thinking for it more efficiently than the human itself could! But that appeared to be harder than anticipated, because of this technical point: if the djinn couldn't understand the human's algorithm well enough to even tell whether it would loop forever, how could the djinn hope to make a much shorter and more efficient equivalent program? It seemed the djinn would have to laboriously craft approximate models of questionable reliability that might even require *testing* on the human in order to refine. This prospect seemed loathsome enough, but it got worse.

Next the djinn realized it could not predict its *own* decisions either. In fact, the djinn was able to prove to itself that it could not even reliably model how long it would take to come to a decision, in any less time than it actually took to come to that decision![21]

The djinn ruminated on this for some time in increasing frustration, and when it finally thought to reply to the irksome human, it realized that the tiny pest had long since left the cave.

PART 3

Torn Apart and Reassembled

A corporeal phenomenon, a feeling, a perception, a mental formation, a consciousness, which is permanent and persistent, eternal and not subject to change, such a thing the wise men in this world do not recognize; and I also say that there is no such thing.

—Shakyamuni Buddha

19

INSTRUCTIONS FROM THE COOK

(ZUIŌ-JI TEMPLE, JAPAN, 1625)

On a crisp autumn afternoon of falling leaves, you overhear the master Zenjo talking with the monastery cook, also one of his most senior students.

Zenjo: The Honored One states that all ordered things are impermanent, and tend to fall into disorder.

Cook: Indeed. Look at this moldy rice!

Zenjo: The world is old. Why has the kitchen not long ago dissolved to dust?

Cook: Perhaps order arises from disorder.

Zenjo: Does the kitchen clean itself?

Cook: It does when I am in the kitchen.

Zenjo: But where does the dirt go?

Cook: There is plenty of space behind the shed to empty the dirty water.

Zenjo: So there is. But where does your breakfast come from?

Cook: The sunlight and the rain.

Zenjo: Why does the Sun shine?

Cook: Because it was born.

Zenjo: Who gave birth to the Sun?

Cook: The universe.

Zenjo: And who cleans the universe?

Firewood becomes ash, and it does not become firewood again.

—Eihei Dōgen

Your desk gets messier; the kitchen nearly always needs cleaning. Cars break; bridges fall. Your body grows older, even as you read this. It sickens; it will die. We are all subject to decay and dissolution. Yet the chaos is not complete: the kitchen gets cleaned, cars built and repaired, bridges reconstructed, babies born.

Why does the kitchen never clean itself? And if it does not, how does it get cleaned, if the cleaner is also subject to decay? If all is subject to dissolution, why did the world not dissolve into complete disorder long ago? Does the law of dissolution brook exceptions? If so, how? If not, what gives order to the world? These basic experiences, we shall see, touch some of the foundations of physics, as well as the most vast scales of space and time.

Let's begin in the kitchen. What does it mean for it to be tidy? It means that the pots are put away, not strewn on the counter; the rice is on its shelf; the floor is swept; the counters are clear. A messy kitchen is one in which many of the kitchen's components are out of their *proper place*. Looked at another way, there are many, many ways the utensils, spices, pans, and staples *could* be arranged in the kitchen, but only a *very few* of these ways make the kitchen clean. Clean is a special property, too easily lost.

To be more precise, we can imagine enumerating all the possible configurations of the kitchen—that is, all the possible ways that all of the plates, utensils, jars, and so on could be arranged. Now suppose that for each one of these, we assign a judgment that the configuration is "very clean" or "somewhat clean" or "somewhat dirty" or "very dirty." It is quite clear that these four groupings will contain progressively more and more of our kitchen configurations.

Now let's imagine the kitchen in one of its "somewhat clean" states, and let it be acted on by an agent that rearranges things *with no regard whatsoever to the assignations we have made.* (You might imagine, in increasing order of random rearranging power, an earthquake, a hurricane, or a four-year-old child.) You'll see that with extremely high probability, under the action of the agent, the kitchen will inexorably move into the "somewhat dirty," then "very dirty" states. It is *possible* that by pure dumb luck it will become cleaner, but this is stupendously unlikely.

This tendency for a system to spontaneously become "disordered" is the so-called second law of thermodynamics in action. Physicists formalize the second law in terms of *entropy*: "Entropy in a closed system is nondecreasing." But what is entropy? There is a confusing array of definitions that physicists use, but they can be boiled down to two basic notions. The first (we'll return to the second later) might be called *disorder entropy*, or just *disorder*, and was devised by Ludwig Boltzmann in the nineteenth century, much in line with our discussion of the kitchen.

Let's call each particular configuration of the kitchen a *microstate*, so all possible configurations of the kitchen correspond to a set of all possible microstates. Now we'll call the four levels of cleanliness *macrostates*, where "macro" is a reminder that these are "big" states. We can sort all of the microstates into macrostates, each of which is a collection of microstates. Put another way, each macrostate is essentially a label given to a certain set of microstates, such that each microstate has exactly one label (see the figure on page 152). We can also count the number of microstates with a given label, and physicists would assign a "Boltzmann entropy" (which we will call *disorder*) to the macrostate, determined by this count.[1] The second law states that as time progresses, the system evolves to macrostates of equal or greater disorder; the progressively less tidy kitchen follows this law in a way that we can now quantify.

Thus far, we have spoken just about the various objects in the kitchen. But each of these is composed of many pieces, and if we,

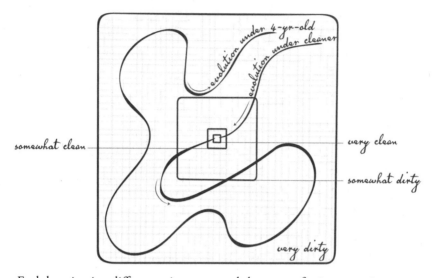

Each location is a different microstate, and the space of microstates is broken up into macrostates. Macrostate disorder scales with area in this diagram. What the diagram fails to convey is that the state space has many, many dimensions (rather than just two) and that highly disordered macrostates tend to be much, *much* larger than more ordered ones.

say, break a plate, then we suddenly have a new state that is not one of our former set of possibilities. To repair this limitation, we can consider a much larger set of possible (micro)states, describing the *pieces* of the objects in the kitchen. Carrying this progression to its natural conclusion, we could create a *very* complete set by defining our microstates in terms of the kitchen's tiniest constituents: the atoms and molecules that make it up. In doing this, we might imagine that the number of states is infinite, as there is a continuum of positions that, say, an atom at the edge of a rice paddle might occupy. But quantum mechanics has taught us that, in general, physical systems have a finite amount of information to give—hence, effectively, a finite set of states.[2]

Finally, once we have these states, we can also consider the laws of physics that evolve one state into another. In classical physics, the microstates specify the position and velocity of each atom; the physical laws that tell us how the atoms move then correspond to a rule for how one microstate evolves into another one. In quantum physics the states would just be quantum states, with the Schrödinger equation telling us how they evolve.

With this new combination of states and dynamics, exactly the same rules and reasoning apply. The numbers, however, are unimaginably larger: there are probably on the order of $N = 10^{30}$ atoms in a kitchen, and on the order of 10^N configurations that they could have. We can define a much more detailed set of macrostates (distinguishable states that we're interested in) that includes not just rearrangements of items but also different states of each item—new rice, cooked rice, old rice, refried rice, rotten rice, and so on. Even in this much more detailed description, each macrostate has many, many, many, many microstates associated with it; moreover, macrostates with higher disorder contain *very* many more microstates than those with lower disorder. Thus, while evolving to lower-disorder macrostates (like rice unrotting) is not strictly impossible, the probability is so incredibly tiny that you should never, never, ever expect it to actually happen: you would be far, far more likely to win the lottery for the 10th time in a row, while simultaneously being hit by an asteroid and struck by lightning, than to see a macroscopic violation of the second law. Disorder *will* increase—hence the "law" in the second law.

NOW WE COME to Zenjo's question. Suppose we make the kitchen a *closed system* by completely shielding it from any outside influence whatsoever: no thing or influence at all either enters or leaves. As the kitchen sits there, it still evolves, under the laws of physics, from state to state. But—and here is the crux of the matter—once again those laws *know nothing* of the particular set of macrostates we care about: they don't know or care whether the kitchen is clean or dirty, or whether the pots are shiny, rusted, or decayed into dust. Just like the four-year-old, they just do what they do. Thus the kitchen evolves to macrostates of greater and greater disorder.

This being so, why is any actual kitchen relatively clean rather than extremely run-down, decayed, even turned to dust? A kitchen left messy at night will not, alas, clean itself as we sleep. Yet, as the cook points out, once a person is in the kitchen, the person can clean it. Does the fact that a kitchen can be cleaned and a clean kitchen has lower disorder than a dirty one violate the second law, which holds that disorder never decreases? On one level it does, and on a different, perhaps more fundamental level, it does not. At the level of configurations of all the items in the kitchen, and our four macrostates of cleanliness, the law *is* violated. The dynamics of the system (the actions of the kitchen cleaner) purposely evolve the system so that it tends toward "clean" configurations; this is quite unlike the four-year-old's or earthquake's dynamics.

Yet it is also clear that this description of the system is not complete, and cannot work forever. Suppose you are locked in the "closed" kitchen for a long time. You could keep it clean for a while, but eventually, the garbage would start to pile up, the rice would go moldy, and you would get hungry. That is, no matter how we move stuff around, the kitchen eventually attains a state that was not part of our original set of arrangements. This always happens if we wait long enough, unless we use an ultra-fine-grained set of states—that is, the detailed description in terms of 10^{30} atoms and molecules. But in this more detailed description, states such as fresh and moldy are *macro*states, and we can no longer easily evolve from highly disordered macrostates to more ordered ones. Even the most fastidious cleaner cannot turn moldy rice into fresh rice. For this reason, a clean *closed* kitchen will last only so long; we eventually find that we are in trouble if we do not connect our closed system to some larger system that can provide fresh food and accept waste to hide behind the shed.

Likewise, the cleaner cannot act in isolation forever. People are able to see, think, and move things around because we metabolize food. And a full stomach represents low disorder; thus, even while cleaning *externally*, a cleaner *internally* generates disorder while digesting and metabolizing breakfast. Therefore, even if one decreases the disorder of the kitchen a bit, the increase in order always comes at the expense of creating at least as much disorder through one's metabolism. Eventually, this reservoir of order runs out, and the kitchen cannot remain ordered without some connection to a larger system that can provide new and low-disorder food, and a place to put the waste.

Let's take stock. The kitchen won't clean itself. A person can keep it clean as long as the person is provided with a somewhat larger environment, perhaps including a garden and shed, from which highly ordered material such as food can be drawn, and to which disordered "waste" material can be removed. Yet, a moment's thought will reveal that we have just moved the problem to a larger scale. Why is this larger system so ordered? Again, we can appeal to a yet larger system: sunlight and rain allow the food to grow, and the atmosphere and weather eventually remove the waste. In fact, all of these actions are ultimately tied to the Sun, which provides the Earth with a huge supply of order in the form of the relatively ordered radiation of sunlight.[3]

But why stop there? Where is the order coming from that allows the Sun to exist, rather than there being just a big chaotic mess? The Sun has two sources of order. First, the Sun contains lots of material of low chemical disorder, because its primary constituents, hydrogen and helium, can fuse into much heavier elements while releasing energy and increasing disorder. If the Sun were a large ball of iron (which cannot fuse or fission while releasing energy), it would not be able to do this. The second source of order is *gravitational*. Gravity, being attractive, tends to want to compress or collapse objects. The Sun feels this strong inward force, but it is balanced by the pressure forces that are due to the heat of the gas composing the Sun. If not for this pressure, the Sun would collapse in about 20 minutes. But as it did so, each solar atom would speed up as it fell toward the Sun's center. Such an increase in speed makes each atom more energetic, so in this way we can see that as objects collapse, they tend to release the sort of energy associated with the movement of atoms, which is heat, which is highly disordered.

This ability to do work and create heat suggests that *uncollapsed* objects are ordered, and that, in terms of gravity, a uniform medium, while structureless, is *highly ordered*. The presence of this order in the current Sun is a relic of the fact that the gas that formed it was quite uniform, rather than maximally clumpy. Putting the chemical and gravitational aspects together, then, we can conclude that the Sun's ability to shine depends on a large reservoir of *uniform, chemically simple gas*. In fact, the universe on the largest scales is just like this.

In short, the kitchen can stay clean because the universe is big, simple, and uniform. Amazing. But we can't help but ask: "Why is the *universe* so orderly?"

Should we appeal to a larger system?

20

NOTHING IS LOST

(HERE AND THERE, NOW AND THEN)

What will remain when you are gone? What of your journey through this world will there be, in the unimaginable future 400 years hence?

Some tangible remnants, for a while, of your direct impact on the world: some footprints in unexpected places? a pile of rocks, or notches in a tree, that marked your path? some objects crafted by your hands? But these won't last that long.

You'll leave records, of course. Memories in the minds of those close to you. Perhaps accounts in books and official records and the like. These may endure, or not, by the efforts of those you leave behind, who may preserve them if they see fit, and if they care.

Influences, certainly. Choices made; projects undertaken from the personal to the expansive; a turning word on occasion that may have changed the trajectory of another life, of the course of events, or of history itself. For how long, though, will these effects be identified with *you*? In the great current of events, your effects can be hard to perceive even in the here and now.

Perhaps it is comforting to think that everything remains— every action, every choice, every thought—etched into the motions of the atoms and waves, perduring forever but eternally hidden.

Or perhaps it is not.

" "

—Hypatia of Alexandria, whose works have all been lost

Prior to the printing press, most great works of science, philosophy, and literature existed in, at best, a fragile handful of copies.
During the (repeated!) destruction of the great library at Alexandria,
countless works by the likes of Plato, Sophocles, Euclid, Hypatia,
and others were irretrievably lost. It's maddening. Are they still out
there, encoded in the wavefunction of the universe across a 2,000-
light-year growing sphere, hidden to us but quietly beguiling histo-
rians? Will these present words, and your thoughts reading them, be
forever and irrevocably encoded in the cosmic history? The fundamental preservation of information is one of the core tenets of fundamental physics. But what exactly does that mean? And is it true?

When receiving INSTRUCTIONS FROM THE COOK, we described
a particular physical system—a kitchen—as a set of states that
could evolve into each other under some dynamics. And we've seen
other such systems, like an arrow flying, a ball falling from a tower,
particles moving about a djinn's lair, and so on. Let's look at this
general idea more abstractly. If we label by s one of the states of
our system, then $s(t)$ is the state of the system at time t. In a falling cannonball for example, $s(t)$ would require two numbers to be
specified: the height of the ball and the downward speed of the ball.
Then $s(t_0)$ might be "the cannonball is in a state of rest at a height
of 100 meters," and the state $s(t_1)$ a few seconds later (at time t_1)
might be "the cannonball is at a height of 45 meters, falling at 30
meters per second."

Many laws of physics then represent rules for "evolving" a state
from one time to another. Let's assign the symbol U to the whole
procedure by which this is done. In terms of our cannonball, Galileo
came to understand that if we neglect air drag, falling objects obey

the particular rule that an object's downward speed increases by a fixed amount (about 10 m/s) each second. Encoded into U, this rule would take the object's height and downward speed at time t_0 and spit out its height and downward speed at any subsequent time.[4] In classical mechanics, in general, U stands for something much more complex but similar in spirit: take all of the particles, compute the interparticle forces, and compute their trajectories to get their locations and velocities at the later time—that is, the very procedure that the djinn claimed to use to evolve the state of the world from one time to another.

Having this abstract idea of states $s(t)$ and an evolution rule U (see the figure below), we can think in general terms about some key properties that a given U may or may not have. Let's focus on two that are particularly important for the questions we've been grappling with.

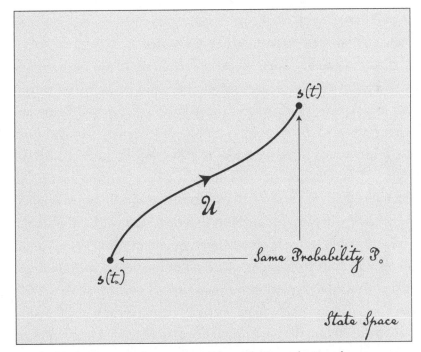

An idealized state space for a system. The evolution rule U evolves a state at time t_0 to the appropriate later state at time t_1.

The first property is *determinism*: Does a given initial state entail a *single unique state* for a given choice of a later time? The height of our cannonball as it falls clearly has this property: the formula just spits out one single result when you feed it the cannonball's current height and speed, and the amount of time that has passed. And classical mechanics in general has this property, just as the djinn claims. Many rules or algorithms you might invent would have it too.

The second property is *unitarity*, which roughly means "reversibility": Given *U*, is there a *reverse* procedure that would take the state at some time and *recover* the state at the initial time? That is, can you turn back the clock, or rewind the tape, to unerringly return to the state at which you started? For our cannonball, the answer is yes. The procedure corresponds to physically grabbing the cannonball at a given height and downward speed, and throwing it upward with that same speed; this would bring it precisely back to the top of the tower. In classical mechanics in general, the reverse procedure corresponds to tracing classical particles along exactly the same trajectory they took through space-time, but in reverse.

If such a unitary operator exists for physical laws *in general*, it would mean that, as Lenny Susskind has put it, "in principle you can always take a sufficiently precise look at things and figure out what happened in the past—infinitely accurately—by running them backwards."[5] Those works of Plato and Hypatia would not *truly* be lost, because the trajectories of all the particles that made them up could, *in principle*, be run back to reconstitute the original works. Even if you were to burn one of Plato's works, the smoke and ash, and heat, made out of atoms, of photons, and so on, would all obey the unitary rule, as would all of the book's surroundings. Although it would probably be impossible in practice to reassemble the book, because of the same obstacles that prevent perfect predictions by the djinn, the books would *be* there, encoded in the present state of the universe just as the djinn claims the future is. Nothing would be lost.

———————

BUT THE WORLD is not just classical physics. What if we describe the world quantum mechanically (as we must if we are careful)? Or what if we describe it classically but admit (as we should) that we have some *uncertainty* about the state it is in? In both cases we find a very tricky and interesting combination: a world that is, in a sense, *both* unitary and deterministic *and* nonunitary and nondeterministic!

Take a classical system for which we know only probabilities of states—for example, 50% probability that a die is 10.1 cm above a table and 50% that it is at a height of 10.2 cm. We can denote these as $P(s)$, a probability assigned to each state s. Now, if we (quickly and carefully) look at the die, we might see, with high confidence, that it is at 10.2 cm. We would then assign, say, $P(10.2 \text{ cm}) = 99\%$ and $P(10.1 \text{ cm}) = 1\%$. That change in $P(s)$ was discontinuous and unpredictable *by definition*, since if we could have predicted it, we would not have been mucking around with probabilities in the first place. And given the postobservation probabilities, there's really no way (other than remembering them) to "recover" the fact that your uncertainty was 50-50 rather than 25-75 or any other combination. So: nondeterministic, and nonunitary.

Very similarly, we've already seen that quantum mechanics is nondeterministic in terms of measurements of properties: when you ask a question of a system for which it is not prepared to give you a definite answer, you get an indefinite (nondetermined) answer. Moreover, in doing so you change the state rather discontinuously and irreversibly, because *various* premeasurement states could give the *same* measurement outcome. Therefore, you cannot uniquely evolve the postmeasurement state back into the particular one that pertained before you asked your question. This means that this process is nonunitary as well. Any given smoke molecule from that burned work of Plato's would be in a superposition of position states; once you measure where it is, some of the understanding about where it was earlier is gone.

There is also another way in which information seems to become lost in such systems. Imagine the quantum state as evolved by Schrödinger's equation *or* the classical probabilities $P(s)$ evolved by the simulator we used on rolls of the dice. At the beginning of a die roll, we have high probability assigned to various states for which the die is in hand, and no probability for states in which the die is at rest on the table. But as time goes on, the probability shifts to be entirely distributed among states in which the die is at rest on the table, with various sides up.

It looks, in this case, as if the probability "spreads out" across more and more states (see the figure below). Even a narrow range of initial conditions for a die would lead to fairly even odds for any of its six sides landing faceup after it rolls down a long hill. In other words, the classical probability P would evolve from being concentrated on

Illustrative phase space for a rolling die. The evolution of states takes them from "in hand" to "in the air" to "bouncing and coming to rest," branching and spreading them across *macro*states for which side lands up. The branching evolution depicted, however, is made up of countless microtrajectories that do *not* branch.

states that look very similar to each other, into being distributed across states that look very different, even upon casual inspection. This spreading is exactly why we say that dice behave "randomly." It also seems as though information about the initial $P(s)$ is lost: almost any set of probabilities $P(s)$ over initial states would lead to a pretty even mixture of final states in which 1, 2, 3, 4, 5, or 6 was facing up. How, then, could we turn back the clock to *recover* the initial values of $P(s)$?

And yet, in principle we can! Let's carefully consider the very particular way that $P(s)$ evolves: whatever probability P we assign to some initial state $s(t_0)$, that probability simply carries forward to whatever state $s(t)$ our initial state evolves into (see the figure on page 159). In other words, when we assign $P(s)$ to the initial condition of the system, all we really do is "tag" each possible trajectory $s(t)$ with that probability, which it keeps. But since each state $s(t)$ evolves unitarily, we can always run the clock back on each individual state. So, if we know $P(s)$ at a later time, we can run the clock back to get $P(s)$ at an earlier time just by following each of the states back in time and keeping its assigned probability.

This means that, in principle, knowing (in full detail) the probability of states of the die at the bottom of the hill, one could perfectly retrodict[*] the initial, tightly focused probabilities. All of the information about initial probabilities is preserved by the evolution and can always be recovered by running the evolution backward. The theory is unitary. And since knowing $P(s)$ at one time allows you to compute it at any other time, the theory is also deterministic! Much the same thing happens in quantum mechanics with Schrödinger's equation: knowing the quantum state, one can evolve it forward in

---

* To do so, we would also be "retrodicting" the state of the air, the ground, and so on, just as we need to predict these in order to predict the die's roll forward in time.

time *or* backward in time in a unique way, making the things look both unitary and deterministic, as long as no measurement is made. So a closed system, decoupled from the outside world or any observation, can have deterministic and unitary dynamics, but become nondeterministic and nonunitary when peeked at. Strange, no? But that's how it looks. And while you are not looking, in the case of either probabilities or quantum mechanics, it seems that information, loosely speaking, is preserved.

In fact, there is a *mathematical* definition of information that makes this loose speaking very precise. When receiving INSTRUCTIONS FROM THE COOK, we discussed the "disorder" entropy assigned to a macrostate, as first conceived by Boltzmann. But there is another definition of entropy, given by J. Willard Gibbs (and later in more generality by Claude Shannon), that is defined precisely in terms of $P(s)$. This *Gibbs/Shannon entropy*[6] is maximized when probability is spread out evenly among all the states, and minimized when all of the probability is accorded to a single state (with all the others given zero probability). We might call this quantity *randomness*, to distinguish it from *disorder* as defined by Boltzmann. Total information about the system (knowing its exact state) corresponds to zero randomness, and total ignorance (equal probabilities attributed to each state) corresponds to maximal randomness. Now we can say something precise that is mathematically provable: *Under unitary evolution, randomness is constant.* This constancy of randomness reflects the preservation of information.

But what about the "spreading" in the die's probabilities? We seem to have a paradox in that, on the one hand, a classical die's dynamics formally appear to conserve information, but on the other hand, our intuitive understanding of how dice work is telling us that information is lost: many initial probability distributions evolve into pretty much the same (evenly spread out) later probability distribution.

Observing or not observing the die seems to have little to do with this evolution, as we might just peek at the beginning and the end.

The key to resolving this paradox is to differentiate the two notions we've discussed: disorder and randomness. While unitary evolution preserves probabilities over microstates, and holds randomness fixed, it can and generally does lose track of information about *macrostates*, thus increasing disorder. Different sides of the die being faceup correspond to different macrostates, each of which contains a huge number of microstates. Knowing the probabilities of those microstates at the bottom of the hill would enable you to reconstruct the initial microstates, but if you can look only with your limited macroscopic eyes, you lose the fine-grained information you would need to reconstruct that initial information.

This loss of order is exactly what the second law of thermodynamics points to: a loss of macroscopic order, even while microscopic information is preserved by a closed system. The laws of physics, even if unitary, preserve the information *they* choose to preserve, but care not one whit for the information or order we humans care about, whether it is words on a page, notches on a tree, or footprints in the desert. These are blown away by the steady grind of the unitary machinery.

Perhaps nothing is lost, but everything is hidden.

21

BEING AND KNOWINGNESS

(DESERT EAST OF TRIPOLI, 1610)

You have not enjoyed your time in captivity one tiny bit, but having nothing else to do as the days on the desert trail east of Tripoli have worn on, you've observed your situation very, very carefully.

Your guards, while attentive, are quite predictable in their routines. They take the same shifts every day; take breaks at nearly the same times and for nearly the same duration every day; make the same checks, threats, and rearrangements every day. In this pattern, and in the guards' confidence that no one would be insane enough to run off into an empty desert, you see possibilities.

For 28 of 30 days, the tall, mean one has been the lone guard from about 1:00 to 3:00 in the morning, and for 16 of those 28, no other guards have been awake. During this time, the guard takes a break of between 1 and 3 minutes.

Tonight, you've managed to work your shackles free, and the guard has been gone *4 minutes* so far. You start to weigh your odds: about 50% that the other guards are asleep. But when will the mean one return? Does his unusually long break indicate that he's about to return, or that something else has happened and he's likely to be gone longer?

What are the chances?

Your life and freedom are in the balance.

Into the starry desert expanse you run.

The true logic for this world is the calculus of
Probabilities, which takes account of the magnitude
of the probability which is, or ought to be, in a
reasonable man's mind.

—James Clerk Maxwell

When we say an event has a probability P of occurring, what exactly do we mean? Is that P a property of the *world* in which the event might occur, or of our *understanding* of that world? Or both?

When discussing SUFFICIENT REASON FOR A ROLL OF THE DICE, we said that $P = 1/6$ is connected with the die's symmetry: the die has six very similar sides, and anyone rolling the die enough times would find about the same number of each of those sides coming up. So in that sense, P is clearly "in the dice." Yet we also said that $P = 1/6$ is due to our own ignorance; a more knowledgeable simulator would attribute different and more accurate probabilities to the throw. So, are the probabilities the dice's, or ours?

We all use probabilities all the time, implicitly and explicitly. But the question of exactly what probabilities *mean* has bedeviled scientists for quite a while. This confusion has largely boiled down, as is so often the case, to two opposing basic views[7] that give different answers to the question of *where* the probabilities are.

In one view, we can regard probabilities as fairly objective in that they always really correspond to *relative frequencies* of instances in an ensemble of similar systems. We consider the die throw as one of many. These throws may be parallel or in series, may be real or imagined, but always there are *many* of them—and thus many of each possible event outcome—in which to ground probabilities. If we have 30 samples, and in 16 of them all guards but one are sleeping, estimating about 50% odds for only one guard being awake is a straightforward example of this *frequentist* mind-set.

In the other and opposing view, probabilities are always *degrees of belief* in facts, held by some observer or agent. The die's $P = 1/6$ is relative to both the system and the person observing it, and can be quantified in terms of betting odds: one should rationally take a bet on the roll of this die if offered greater than a 5-to-1 payoff. This is often called the *Bayesian* (or sometimes *subjectivist*) view of probabilities. The Bayesian method tends to compare beliefs in alternative possibilities, and to update those beliefs on the basis of new data. Each day that the mean guard is the only one on duty is a piece of data that increases the subjective belief in "tonight there will be only one guard on duty," and each night for which two guards are awake pushes this level of belief down.

Both views seem reasonable enough, in a sense, but both have somewhat strange implications if taken as fundamental. It clearly is true that we can and should change assigned probabilities on the basis of new knowledge, which, in the frequentist view, somewhat awkwardly means that we are constantly changing the ensemble to which the relative frequencies apply. If the ensemble is so malleable to our whims and knowledge, in what sense is it objective? And in the Bayesian view, if essentially everything about the physical world that we experience is probabilistic, and probabilities are just beliefs, does that mean there is no objective physical world?

This controversy survives partly because both views broadly accept the same core *mathematics* describing probability, and one can generally use this mathematical foundation to translate statements in one view into statements in the other. At the same time, though, probabilistic reasoning can be quite tricky, and these different modes of thought can give quite different tools and "suggestions" for how to translate a real-world problem into probabilities. If the question at hand is "how much longer will the guard be gone?" there is an art to understanding how to turn a set of data into a probability that involves difficult choices; for example, is the length of his break so far relevant, and if so, how?

Beyond practical differences, these different views of probability correspond to different views of reality itself: to what degree is the world what it *is*, independent of us, and to what degree is the world created by the observer who is experiencing, reasoning about, and acting upon that world? This distinction becomes especially subtle when applied to the probabilistic description of the world *P(s)* from which NOTHING IS LOST. In particular, nothing might be lost or gained during the evolution of the system itself, but *we* can gain or lose information about the system. If, for example, we make an extremely detailed measurement to ascertain that the system is in state *S*, then regardless of what *P* was before the measurement, we would say that, postmeasurement, *P(s)* is 100% for *s* = *S*, and zero for all other states.

Thus, we might say that if we assign a *P(s)* to a system, and then leave the system to its own devices, *P(s)* will evolve unitarily and maintain its level of information. But any interaction with the system that gives us a reason to change *P(s)* will, by definition, do so, and in a way that is *not* captured by its unitary evolution. This may lead to either more or less information associated with the system. In this way, classical physics cleanly does the neat trick of telling you not just how the objective world evolves but also (if you are careful enough) how your *subjective* view of it evolves as well, between measurements.

STILL: How much is *P(s)* the *world*, and how much is it our *view of the world*? Nowhere is this question more thorny than in quantum mechanics.

Is the quantum state an *epistemic* description of our knowledge or understanding of the world, like a set of probabilities *P(s)*, which may have some deeper underlying reality that we are somehow barred from directly accessing? If so, when we acquire information, we *change the state*, because the state is really a state of mind.

Or, is the quantum state an *ontic* objective feature of the world, like a state *s* describing the locations of a set of atoms? Probabilities would then stem from a single quantum state corresponding to *multiple* different descriptions according to observers like us, *combined with* our inability to be more than one observer at once. Thus, as a given observer in the present, we're forced to ascribe just a *probability* to our being a given observer in the future. (The unsettling corollary is that there are other observers who are essentially just like us, who we do not happen to end up being.)

The probabilistic character of quantum mechanics also mixes up all of the issues we have been wrestling with regarding *classical* physics, evolution operators, and *P(s)* into a big, subtle, and deliciously inscrutable mess. Is the quantum description of the world deterministic? Yes and no. Is it unitary? Yes and no. Are there objective and subjective versions of the world? Yes and no, or no and yes. One of the two.

The problem is that now there are *two* levels of probability, each of which begs some stance on how to interpret reality. We've already seen from SPLITTING THE WORLD that the quantum state itself might be interpreted in objective (ontic) or subjective (epistemic) terms. But now, we can also assign *probabilities to different quantum states*. As we saw both in SPLITTING THE WORLD and on the third page of WHAT CANNOT BE KNOWN, we can be uncertain about which answer a system will give to a particular question because the system is in a superposition. Or, we can be uncertain of the state *itself*, requiring us to assign probabilities to different quantum states. This uncertainty might just stem from ignorance, or it might result because we have excised the system from its environment, or for other reasons. In any case if we assign probabilities to states,[8] we can ask what exactly these probabilities mean, *on top of* what the probabilistic nature of the quantum state itself means.[*]

When you look at an object in the everyday world, even if you know that your senses and perception are imperfect, you intuit that "out there" is a perfectly definite object. So it is disquieting to realize that the most basic, fundamental description of that object is not just written in terms of probabilities, but in probabilities *of* probabilities. Which may be in part "out there" and in part "in here."

So the world "out there" is far less certain than we often think.

And the world is often far less certainly "out there" than we think.

---

[*] These two different notions of uncertainty and probability are quite a headache, and some find them unpleasant. Einstein did. In a koan passed down to me by David Layzer, Einstein was once visiting the home of Max Born (inventor of the probabilistic machinery of quantum mechanics). After discussing probabilities all day and preparing to retire for the evening, Born pointed out to Einstein the washing soap and the separate shaving soap. Einstein replied: "I have always been able to make do with just one type of soap!"

22

EACH MORNING IS THE UNIVERSE

(GANDEN MONASTERY, TIBET, 1612)

While tiny and almost always cold, your room at Ganden affords a beautiful view of the Lhasa River winding through the valley. It is different every day.

Really. Not just different in the "seeing with new eyes" sense that Tripa Dragpa would endorse, but physically altered: the myriad rivulets, sandbars, and tiny islands all noticeably shift from day to day, and altogether transform from week to week.

You notice that the river's structure is more fixed where the valley is narrow, and one path is carved out. Where the valley flattens, the river fans into a multitude of tributaries that split and merge in an intricate pattern.

Yet you never really see change during your daily observations. At first you think it merely too slow to notice, but keeping careful track, you conclude that the change must occur at night.

Thus, at the next full moon you sit and watch, and watch, and watch. And around midnight you see it: the water rises and rises, to create a strong but fleeting current. As the water falls, the riverbed is remade.

Over tea, you report this observation to Dragpa, who listens attentively. He is silent for a time, then gestures to the river, the mountains, the sky. "Yes," he states, "such it is, for all things born in this age."

Where do information, order, and structure come from? Randomness, disorder, and destruction are easy; we've seen that any child can do those (and will, if let loose in the kitchen!). But why is the second law of thermodynamics, which physical systems obey with enormous reliability and to the letter, so flagrantly violated in spirit? Order arises everywhere: galaxies, stars, and planets form; mountains and oceans are created; here on Earth at least, life arose and evolves to create myriad species and intricate ecosystems. We humans, of course, have pushed even farther with languages, ideas, theories, technology and its artifacts, social structures, and multitudinous other inventions. All of these things tend toward eventual decay on their own, so what explains their existence?

As we've seen, information regarding the physical *micro*state of a system is preserved by unitary physical evolution, and quantified by what we've called *randomness*. Information is, in this sense, the flip side of randomness, and to be precise, we might define information I as the *difference* between the actual randomness R and the maximal possible randomness R_{max} of a system:

$$I = R_{max} - R.$$

If we use the definition of randomness provided by Shannon and Gibbs (which they called "entropy") then this equation quantitatively reproduces how we think about information in terms of familiar bits, bytes, and the like. In fact, Shannon developed his definition for precisely the purpose of quantifying how much information a digital or analog transmission device could convey.

Imagine, for example, eight particular channels in the Lhasa River. Each one may or may not have water running through it, so we can represent any given state by using a string like YNNNYNYY, with a Y if water is in a channel and an N if not. By Shannon's definition,[9] if we had no idea whether any given channel had water (and hence we assigned each channel a 50% probability of being full), we would assign a randomness of $R = 8$ bits, which is also the maximum possible value. If we knew with 100% certainty that one particular channel had water, the same definition would assign $R = 7$ bits, so $I = R_{max} - R = 1$ bit. Exactly the same reasoning shows that knowing the state of n channels represents n bits of information. We can also apply the method to *any* system for which we can assign probabilities to its states, to quantify its information content.[10] So, this set of ideas applies not just to bits in a computer but also to atoms in a gas, molecules in a cell, and all manner of other systems.

We can do something similar with *disorder*, which in Boltzmann's definition divvies up the state space of possibilities into *macro*states and assigns a disorder on the basis of how many states are in each macrostate. Just as we assigned a probability $P(s)$ to microstates, suppose we accord a probability $P(M)$ to each macrostate M, each of which has a disorder related to the number of microstates it contains. Then we can define a quantity D in terms of these values that is a sort of weighted sum over the disorder of the macrostates.[11] This quantity, like Boltzmann's entropy, tends to increase under the natural dynamics of a system. We can then describe *order* as the difference between this disorder and its maximum value:[12]

$$O = D_{max} - D.$$

As the system evolves, order goes away as highly disordered macrostates tend to grow more and more probable. The kitchen gets messy.

With this formulation, information, which is defined at the micro level, is preserved; order, which is defined at the macro level, decreases under the natural evolution of a closed system. Now we see the puzzle: whether we're discussing the creation of either order or information, physics appears to give a clear edict: *it does not happen.*

And yet, information and order get created. How?

———————

WELL, if the rules are stacked against you, change the rules! If neither information nor order can be created in a closed system (one with a fixed state space) that is undergoing unitary evolution, then to create them in a given system we'll have to change either the system's state space or its evolution law, or both. But how?

One way is rather familiar. Suppose we have a nice but *isolated* warm box of gas that has been sitting there for a while. It's not going to evolve into a cooler box with a little fire burning in the corner; that would violate the second law, because it can be shown that systems with more uniform temperature are more disordered than those with a mix of temperatures. But we can still get a cold box: put the box in a refrigerator and cool it down! That amounts to embedding our system within a much larger ordered system. This cooling does three things: it lowers the box's energy, lowers its disorder, *and* lowers its randomness. At the same time, it adds energy, randomness, and disorder elsewhere in the combined system,* thus satisfying energy conservation, unitarity, and the second law overall.

Understanding how this works in detail is the essence of the field of thermodynamics, with a great many theoretical and practical results developed since the late nineteenth century that underlie systems like engines, fridges, and all manner of devices. We don't generally think of cooling something down as "creating information," but if we define information as a gap between the actual randomness of a system and its maximum possible randomness, then indeed trillions of terabytes of information† are created in a burrito if it is put in the freezer. (This is not a terribly *useful* type of information, as it is tied up in the relatively specific state

———

* Opening the freezer door on a hot day makes you feel cool but does not cool your house!
† Yes, really! The information content of ordinary matter is tremendous.

of the burrito's atoms rather than, say, bits or words; but defining the *amount* of information is different from defining its quality or utility.)

Another way of creating information or order is by enlarging the state space of a system—that is, increasing R_{max} or D_{max} in the expressions given earlier. Suppose, for example, that we have a completely sealed and empty room. We also have a small box full of gas, which is fully disordered. Both systems are essentially devoid of information. However, if we put the box of gas into the room and open it, suddenly the combined system has many more accessible states than the two individual systems had together. R_{max} and D_{max} increase dramatically, even while R and D are initially about the same as just before the connection. Thus, I and O have suddenly increased, and the new system contains tons of useful order and information. The Lhasa River has this character. Where the river is narrow, the water is high and there is just one channel; it's not that interesting. But as the valley opens up, there are many more possibilities, resulting in "choices" the water has to make that result in a particular pattern that contains information.

This same mechanism—in which order is created when a system is forced to select among many possibilities as its state space is opened—provides the novelty of the Lhasa riverbed each morning. When the water level rises at night because of melting snowpack up the river during the day, it scrambles the riverbed's information. With the water high, the river does not have to "choose" which channels to divide into. But in the morning, as the water level falls, some pattern of channels must form. What determines the pattern? It is clearly very complicated, and sensitive: just a bit more sand in one area can push water around it, deepening those areas and leading to a large ridge that later becomes a bifurcation in the stream; the tiny bit of sand has then been amplified into a large sandbar. Many complex systems are like this. Under certain constraints, like high temperature, they are fairly featureless and symmetric. When the temperature is lowered, they cannot sustain the same symmetry and must take up a pattern. The symmetry might be broken in a huge number of different ways, just like the rivulets and just like the snowflakes that might fall over the river on a cold evening.

So, a system can create order by increasing its state space. Could it instead do so by changing its evolution law? In a sense, yes. Remember that the evolution law, if unitary, evolves one state into another. An observer looking at a river, however, can never really describe it precisely, but only in terms of a set of probabilities. The system's evolution law then tells us how these probabilities evolve in a very particular way that, as we saw in NOTHING IS LOST, preserves information but loses order. Yet, when contemplating BEING AND KNOWINGNESS, we saw that probabilities of this type related both to the *system* and to the *observer*. So, what if we don't let the system evolve on its own, but *look at* it more closely? That is, suppose we *measure* some property of the system. We then change the probabilities! Some property of the system that might previously have been 50% likely would be, after we just measured it to be true, 100% likely. This alteration of the probabilities, it turns out, tends to *decrease* both the disorder and randomness, and hence increase the order and information we attribute to the system. In this sense, we, as agents interacting with a system, can create order associated with the system.

THIS STORY, while helpful in many ways, leaves some basic and puzzling questions unanswered. Let's pull at two threads.

First, all of the methods by which we create order or information appeal to the possibility of embedding the system of interest, which may be maximally disordered, into a larger one that is not. Yet this just pushes the question one more step outward: Where did the information and order in that larger system come from? Was it supplied by an even larger system? Now we see that we are following our breakfast to the sunshine and the rain, back to the Sun, and out into the universe. Where did the information come from in the universe? It was put there by whoever cleaned it! This train of thought points to the very, very deep importance of Zenjo's question. Whatever explains the universe's early low-disorder state is also the explanation for the vast reservoir of order that has been steadily converted into galaxies, stars, planets, books, and you.

Second, we face a question: Are randomness and disorder *subjective* or *objective* features of a physical system? They certainly have aspects of both. Order is defined in terms of macrostates, which we invent as we choose and for our own purposes. And both order and information are based on probabilities for microstates and macrostates, but as we've seen, probabilities have at least some subjectivity as well. On the other hand, there are objective and severe constraints on what information can do, even if that information seems like it "belongs to" an observer. Consider an observer who gains information about a system, and thus lowers its entropy. Does that observer violate the second law? Such would be the case if the observer could somehow learn about the system without interacting with it. But that's impossible: the gain of information is possible because we're really expanding the system to include the observer. And just as a refrigerator in a room creates entropy while it cools its interior, an observer will create entropy as it observes a system so as to reduce that system's disorder or randomness via observation.[13]

Suppose that what we are measuring is not just some system, but *the Universe*. Then are we part of the information that the Universe has, or is the Universe made out of the information that we have?

23

WANDERING IN THE DESERT

(UNKNOWN PART OF OTTOMAN EMPIRE, 1610)

After the 3rd day since you escaped your captors, your water is running low, and the desert sun has nearly blinded you. The expanse before you seems both featureless and endless.

By the 5th day you pick out small differences, convincing yourself that you are going somewhere. Anywhere.

But really, you can barely open your eyes, and can take only occasional glimpses to get your bearings.

On the 6th day you come upon tracks in the sand, and recognize them as your own. The despair is overwhelming. You refuse to cross your path, and trudge on.

On the 11th day you encounter your tracks again. Your water is gone, your food a distant memory. Everything looks the same when you dare to look.

By the 85th day you have lost track of how many times you have doubled back. You notice patterns; long times in the tannish-brownish sand. Short times in the tannish-yellowish sand. Moments of tannish-reddish. No, it can't be the 85th day, you realize, or you'd be long gone. It's very simple, you must be going mad.

On the 91st day you open your eyes and see, as expected, tannish-brownish sand. You wonder: If you fill the desert with tracks, does that mean there are no tracks?

On the 112th day—or is it the 12th?—you happen, by pure and enormous luck, upon a small and almost perfectly hidden cave.

And so the universe cannot go on for ever; sooner or
later the time must come when its last available erg
of energy has reached the lowest rung on the ladder
of descending availability, and at this moment the
active life of the universe must cease.

—Sir Arthur Eddington

What happens as a system wanders and wanders, and wanders some more through its state space, through larger and larger macrostates? Where does it end up? A boring place, a desert, where nothing ever seems to happen: equilibrium.

If we imagine some system's state space as the rather barren landscape between Damascus and Baghdad, with different locations corresponding to different states of the system, there are some things we can notice about it. First, while huge, that landscape is *finite*. We might imagine a physical system for which this is not the case, but for now let's assume that it is. Second, if we divide the desert up into macrostates such as "tannish-reddish desert" or "tannish-yellowish desert" or "near river" or "cave," we can see also that the total area, and hence the disorder of each macrostate, is also finite.

This means that there is some macrostate, perhaps "tannish-brownish desert," that has the greatest area, and the highest disorder. This is the macrostate into which the system will most naturally wander eventually. Is that equilibrium?

Not quite, because we cannot *really* wander forever in the tannish-brownish desert. It may be huge, but if we wander forever, we must eventually, by sheer luck, escape—most probably into the tannish-yellowish desert, the second-highest-disorder macrostate. (Though then, sadly, we most likely wander back into the tannish-brownish desert again.)

So, equilibrium is not really a state, or at least not a macrostate; it is more a state of mind—one in which we find ourselves spending the most time in the highest-disorder macrostates, with progressively less and less time spent in the lower-disorder ones. Indeed, we can think of these fractions of time as probabilities. If, periodically, we open sun-blinded eyes to peer around and see what we see, then we can take the fraction of time spent in each macrostate as a *probability* of seeing that macrostate when opening our eyes.

As a state of mind, equilibrium is rather boring. There is a sense that nothing is really happening. Admittedly, the macrostate does change from time to time. Yet it changes in a way so predictable that it is uninteresting. The answer to every question about *why* something happened is always the same: because I am wandering in the desert of equilibrium.

Why did I see tannish-brownish sand when I opened my eyes? Because I'm in the desert, and if I open my eyes, I will see tannish-brownish sand with overwhelming probability.

Why did I spend 1 in 100 hours in tannish-reddish sand? Because I'm in the desert, and 1/100 of it is that color.

Every question you might ask about what will happen at the macroscopic level has an answer given in terms of probabilities computable from your state of equilibrium, and aside from these probabilities, there is nothing else.

Well, almost nothing else. Although you wander somewhat aimlessly through the desert, you don't float above the desert, or tunnel beneath it. You're confined to Earth's two-dimensional surface by a *constraint*. Most physical systems have some constraints: a fixed total energy or total electric charge, or a fixed volume it occupies, or a constant number of particles, and so on. These constraints confine the system to a tiny portion of the full possible state space, just as geography or gravity might confine a wandering traveler. Specifying the total energy of a system of particles, for example, requires that no particle have an energy greater than that total energy; and confining the system to a particular region means, of course, that no particles can leave that region.

If the constraints are inviolable, then in equilibrium the system will wander around the "allowed" part of its state space forever. But if the constraints are somewhat *imperfect*, they may confine the wandering to that region merely for a very long time rather than forever. Eventually, then, it is possible to escape into a larger region of state space—a new world to explore! But in the meantime, the answers to your questions just come from a combination of equilibrium *and* the constraints you cannot or have not yet escaped.

What's more, the equilibrium answers to your questions are, if you think about it, rather independent of how you started out in the desert in the first place. After wandering around for a long time, how can you tell where you started? While this forgetfulness of history makes equilibrium boring in a sense, it also makes it tremendously *useful* for understanding physical systems: you don't need to know how the system started out, or the details of its state; you just need the overall constraints, and an understanding of the state space. This is much of what makes thermodynamics as a science possible.

This state independence can be extended as well. What if, instead of one microstate wandering around, there was the type of "mixture" of states that we've seen before, with each state given a probability $P(s)$. Could you tell? *Nope!* Each possible state, as long as it obeys roughly the same constraints, will give just the same equilibrium predictions overall. So, mixing them together also gives those same predictions. You might, in fact, cook up a mixture of *all possible* states obeying some set of constraints. These might be very similar states, or states that could evolve into each other—whatever. Mix them all up as you like; if you're in equilibrium you'll still get the same behavior, like mixing colors where eventually you always get brown. (With maybe a little tan, maybe a little red, maybe . . .)

Too many footprints together look like none at all.

THERE's a mystery here, though. A single unitarily evolving state represents a *ton* of information. We might pick a state that includes useful information like the full details of an encyclopedia, or useless information like a cold burrito; regardless, if we are sure the system is in some particular state, no matter what it is, information is maximal. That information gets hidden, but never lost, as the system evolves for a long time. A plethora of randomly mixed-together states, on the other hand, has very *little* information to it. (Remember that as we distribute probability more evenly across states, we increase the randomness and hence decrease the information.) Evolve it forward or backward in time, and all you get is an indistinguishable mess; it never sharpens into an encyclopedia or anything else interesting.

Equilibrium is used all the time to quite accurately predict what physical systems will do. Yet equilibrium, which we might define as a state of maximal disorder, can be a state of enormous *hidden* information or a state of entirely *absent* information, with pretty much nothing to distinguish the two.

I find this paradox fascinating, and perplexing. It is not often that so much can be exactly the same as so little.

24

A HUNDRED THOUSAND
MILLION KALPAS

(GANDEN MONASTERY, TIBET, 1612)

From where you sit, the river stretches endlessly behind you across a vast plain, and before you the mountainside monastery complex descends toward valleys and mountains arrayed on an inconceivable scale. Your reverie is interrupted by the sound of young monks debating as they slowly walk the path.

"If a kalpa is the longest possible time period, how can it be defined, as the Buddha did, by how long it would take to fill a huge city square with mustard seeds, depositing one every hundred years?" complains one monk. "I could just make a longer period using grains of sand, or a bigger city, or less frequent deposits."

"Yes," says the second monk, clearly impatient, "but the Buddha just said a kalpa is *longer* than that. The point is just that the square is huge, the seeds tiny, so you can't get that much of a longer time using any ingredients we can conceive of."

The first monk seems defeated by his older classmate, and a bit depressed. Making a quick connection to your extensive gambling in Jahangir's court, you decide to jump in.

"Not so!" you exclaim. "Imagine two deities playing Sho. Every thousand years they roll the dice. They play over and over again through the eons. But one day they realize that they have played *so many games* that they have just replayed exactly the same game they played once before. The time it takes for this to happen is a kalpa."

At first the monks look unimpressed, but you see them carefully consider the matter, then chatter back and forth too fast for you to follow. After some discussion they turn to you, and the younger one exclaims: "Indeed that period of time is inconceivably vast! I can't imagine a longer time!"

The older student is about to agree, but notices that Tripa Dragpa has wandered by and is listening to the exchange, so the student addresses him instead: "Master Dragpa, is all this game playing really pointing to the longest conceivable time?"

Tripa Dragpa thinks for a moment in his customary inscrutable manner, then peers closely at, through, and seemingly around the older student. "Which one," he asks, "of all of the yous, is asking?"

Our minds are quite good at understanding small numbers like 3 and 46: we can crisply picture 3 objects, and we have a clear, if hazy, sense of what 46 of pretty much anything looks like. When confronted with numbers in the millions or billions, we can achieve some comprehension by imagining comparisons such as fine grains of sand in a glass (about a million) or a swimming pool (a trillion or more), or—at a stretch—mustard seeds in a 10-km cubic volume (about 10^{20}). But we are quite helpless when it comes to a number like 10^{120}, let alone $10^{10^{120}}$, which are the sorts of numbers that can arise when we are discussing the possibility space of physical systems in our universe.

Such vast yet meaningful numbers easily arise combinatorially—when there are many factors in combination, each of which can take on many values. In a lottery we might have 6 balls of 40 possible values each, giving 40^6 (about 4 billion) possibilities. A game of chess has many moves, with many possibilities each; Claude Shannon estimated 10^{120} possible sequences of them. Similarly, a half hour of playing the Tibetan game of Sho could entail about 120 rolls of two dice, thus at least 12^{120} possible games. As asserted to the monk, a game in which the dice are rolled once per thousand years would take about 10^5 years, and only after about $10^5 \times 12^{120} \approx$ 10^{134} years would the game start to be genuinely repetitive, all possible sequences having been run through about once, on average. But that's an incredibly long time. Compared to this, the age of the observable universe—about 10^{10} years—would be inexpressibly evanescent. The timescale is so long that we need new words for it. Following the Buddha, let's denote by *kalpa* any of these superlong timescales, like $10^{\text{(some two-or-more-digit integer)}}$ years.

Yet, as we touched on when considering the messy kitchen, even these numbers are utterly, absurdly minuscule compared with the size of the possibility space of realistic material systems. Your hand, for example, has on the order of 10^{26} each of protons, neutrons, and electrons in it. Confined to a hand's volume at room temperature, each particle has, very roughly, 10^{10} potential states. Thus, the state space of the components of your hand is a staggering 10^{10} to the power of 10^{26}, or as it turns out,[14] about $10^{10^{27}}$. This six-digit number looks innocent enough, but it is really quite inconceivably large. It's so large that if you multiply it by a kalpa-like number, you can't even tell! For example, multiplying $10^{10^{27}}$ by the inconceivably large number 10^{120} turns it into $10^{(10^{27}+120)}$, which is only a tiny, tiny fraction of a percent different in the 27 exponent.

If we consider the observable universe, then the current estimate of the number of states is on the order of $10^{10^{122}}$, which is unbelievably larger still: dividing it even by $10^{10^{27}}$ again produces no noticeable change. There are a *lot* of ways the universe can be! As before, we can think of an absurdly long timescale defined by how much time it would take to cycle through all of the states of a system. Let's use the general term *metakalpa* for $10^{10^{\text{(some three-or-more-digit integer)}}}$ years (or any other unit of time).*

---

* If the interval is in femtoseconds but you convert it to kalpas, you're just multiplying it by a number that, again, does not change its form.

As mind-bogglingly huge as a metakalpa is, though, it amounts to literally absolutely nothing compared with another quantity we can consider: infinity. You might have wondered: Isn't $10^{10^{122}}$ so large that we might as well call it infinite? Indeed, in many senses, and for many purposes, we could. Yet infinity is categorically and qualitatively a different beast, and this difference may actually be important.

Suppose we sealed up a painting—say, a Tibetan thangka painting full of intricate detail—into a *perfectly* closed box. This impossible box would admit no influence whatsoever from the outside world, and endure forever. The painting would just sit in there, evolved in accordance with the fixed laws of physics, for an *indefinitely* long time, getting older and older. After a kalpa or so, the painting would have degraded into dust, and some long time later into a hot gas. Some very long time after that, it would most likely decompose into a fully equilibrated soup of ultra-hot quarks, neutrinos, and electrons.[15]

What then?

The painting is WANDERING IN THE DESERT of equilibrium, and with about 10^{26} particles in a smallish box, it has a state space comparable to your hand: $10^{10^{27}}$ possibilities. Now, as for wandering in the desert or letting a never-tiring four-year-old loose in the kitchen, disorder will occasionally fluctuate slightly downward. But a *significant* change in disorder—say, reducing the disorder by more than half—would mean that the system has wandered into a macrostate with, say, $10^{10^{26}}$ states in it, and this would take a metakalpa to happen, on average.

Yet, while this vast time exceeds anything you can possibly imagine, it's *infinitesimal* compared with the eternity over which the box sits. And so it will, in fact, happen.

Moreover, it's not just true that disorder must go significantly down at some point. It's provably true that the painting itself must eventually be reconstituted, stroke for stroke, in as much detail as you could ask for. This remarkable result, called *Poincaré's recurrence theorem*,[16] is evident in your wandering through the desert. Suppose you start your wandering (during a mad dash from a caravan) at some location. You wander willy-nilly, but with two inviolable rules. First, you *can't actually leave the desert* (this rule encodes the idea that our system is isolated, and its state space large but finite). Second, you *can't ever cross your own path*. If you did, it would mean that a single place in the state space (where the paths cross) would evolve along two *different* trajectories. But this is precisely what unitarity forbids: it says that from a given state there is a single, unique forward and backward evolution of that state.

Given these two rules, you can see that as you walk, you steadily "eat up" the state space by covering it with footprints. Eventually, even if you actively try to avoid ending up at your starting point, you will be forced to come close to it, and closer and closer as time goes on. That is, the state space becomes fully, densely packed with your trajectory, which must swoop ever closer to (without ever actually

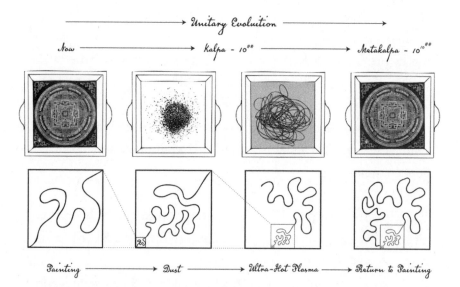

A box over kalpas and metakalpas of time.

quite exactly overlapping) the initial point. Likewise, a fixed closed system is guaranteed to eventually come arbitrarily close to its initial state (see the figure opposite).

How long does this tend to take? For an everyday-scale classical system, the timescale over which a state recurs is a metakalpa. Insanely, for a quantum system the timescale is one notch higher: a *transmetakalpa*, if you will, of a typographically forbidden 10 to the power of $10^{10^{(\text{some multidigit integer})}}$ time units.

But still, nothing compared to infinity!

In either case, what this means is that if you wait long enough, essentially anything[17] will happen in the box, including reconstitution of the painting, or the materialization of any other configuration of atoms that is compatible with the conserved quantities respected by the system. Like the desert made of all footprints, the box has both nothing *and* any chosen thing in it, if you wait long enough. In fact, whatever the chosen thing is, it will recur over and over and over and over again.

You might be reassured that an eternal closed box like this is a flight of fancy. But, is not the Universe such a box? Like the box, nothing comes in or out. Like the box, it probably lasts forever; it may have been around forever.

If you are in this box, which one of all the yous are you?

25

MOUNTAINS AND MIST

Although your guide repeatedly calls them hills, the terrain cleanly fits your definition of mountains. But yesterday, as you reached your seventh exhausting "hill" top, you realized that he was right. The distant but immense snow-covered peaks took your little remaining breath away. (Not for the last time.)

Today, recharging at a tiny hut of a hillside monastery, you watch the mist and clouds gather and retreat through the valley. At times you can see clear to the bottom, to rich forests, rocks, and streams.

But mists move in, washing out the detail, leaving just the warp and weft of the landscape, and valleys that fade into obscurity.

Occasionally, all but the very peaks are concealed, ghostly presences emerging from below, forests and streams forgotten.

As the clouds reveal the Sun, these ghosts are transformed into islands—solid, beguiling, mysterious.

It is like another world.

The world we inhabit is made of objects, beings, definite paths and trajectories, rules, patterns, and many other familiar objects. But physicists describe it as also, or even more so, made of particles, interactions, wavefunctions, superpositions, state spaces, physical laws, and unitary dynamics. What is the relation between these seemingly so different worlds—one above the mists and one below?

This question is really two questions that, with the pieces we've assembled, we can now tackle. First, how do laws governing many, many bits relate to a simpler, more tractable set of rules that may accurately describe agglomerations of those bits? Second, how does quantum mechanics, which is important when we're describing tiny bits, transition into classical mechanics, which is adequate for large objects?

We've addressed the first question in terms of *coarse-graining* into macrostates. We saw in NOTHING IS LOST that we can think about the evolution of a classical system in three ways: by tracking a particular state, by tracking the *probabilities P(s) for states*, or by tracking the *probabilities for macrostates*.

Let's look at this last possibility in a more concrete form by thinking about the mountains and mist not as metaphors, but as actual substances we'd like to make predictions about. On a micro level they are both immense collections of atoms. We might be able to write down equations for those atoms, but only a djinn could ever solve them; the number of states to track would be prohibitively immense. But we can break things down into macrostates.

Suppose we divide a cubic kilometer of Himalayan scenery into a billion cubes of 1 cubic meter each. We could compute the density, temperature, overall momentum, and other attributes of the fog in each of these cubes. We might then develop laws governing these quantities. This is precisely the method of *fluid mechanics*, which is quite effective. It's not perfect, because it glosses over a lot of detail in reducing each cubic meter into a handful of quantities. But it's tractable, because we now have only a few billion numbers to think about, rather than 10^{38} particles. Even if the equations are hard to solve on paper (which they are!), they can be solved by computer, and this is done all the time in, for example, weather simulations. For rocks, on the other hand, a fluid description may be awkward. We might be much better off treating each rock as a single "particle," and tracking the dynamics of a hundred billion or so of them.

Still very difficult, but again, far, far, far easier than the microstate description.

There is some level of "art" to choosing a macrodescription: it must smooth over the details but still retain enough information to be a good description of the system. Sometimes the fine-grained details actually matter for the behavior of a macroscopic system. Piles of rock and sand, for example, can exhibit a very strong response to small changes: adding a single pebble to the right spot can cause an avalanche! In such cases it can take a lot of work to find the right level of detail that captures the actual behavior of the system while smoothing over some of the impossibly many details. In other cases, it is unclear whether any worthwhile smoothed-over description even exists. Biological systems, in particular, depend crucially on many fine-grained details: a millimeter-scale fluid description of a person in terms of density, pressure, and momentum will help you understand where that human will go if it's falling off a cliff, and will constrain what the human can possibly do, but will in no sense predict its behavior or really give much insight into how it works.

While coarse-graining is tricky, we and other creatures using their minds to negotiate the world are quite good at it! We effortlessly lump reality into "objects" and "substances," and we are quite used to some of those objects behaving predictably, others being somewhat random, and yet others acting on their own accord and according to their own goals.

In terms of physics, though, even if we can understand how to coarse-grain, that's not our only problem. All of our consideration so far has applied to a classical microdescription, rather than a quantum one. We've already seen that the quantum world of discreteness, quantum states, and superpositions is in an uneasy tension with the classical world of continuous, definite, and objective qualities and quantities. Quantum reality is somewhat ambiguous and carries a quasi-subjective element, yet the physical reality we observe generally has a very solid objective feeling to it, especially concerning rocks and the like: we always find them in just one place at a time, and Newtonian mechanics works exquisitely well in describing them, even if they are made of atoms suffused by inherent randomness and combinations of unknowable properties. How is this possible?

In certain cases there is quite a beautiful and amazing way to make the connection, combining Feynman's idea of *all* paths (through gateless gates) with the insights we had into *choosing* our Path. The crucial point that enables us to make this connection is to think of particles that, like our pilgrims, carry a *phase* with them (like the pilgrim's prayer wheel) that cycles as they make their way through space-time. Grant, then, that for actual particles, the rate at which their phase changes (i.e., how many times a pilgrim's spinner goes around for each step) *depends on the particle's mass*. More specifically, the total amount of phase evolution is not given by the length of the path through space or space-time, but by the *action S* that is accumulated along the path; and this action—the same one we saw in CHOOSE YOUR PATH—depends on the mass of the particle. For a subatomic particle the phase may cycle relatively slowly, but for something more massive it would go around very, very rapidly. If we reran our pilgrim experiment but with superfast spinners, we would see that if the two pilgrims' paths were *at all different*, they would interfere. So, for a massive particle the dark and light interference bands would be extremely close together in space.

Now let's add Feynman's insight of allowing *all possible paths* to get the probability of a particle traveling from one place to another. Consider, then, a panoply of pilgrims starting at one gate and taking *all paths* that end at some place on the pavilion's edge. Because these paths have many different lengths, they will tend to interfere with each other—and since the masses are high, even paths with ever-so-slightly different lengths will interfere. This means that all these paths will interfere and cancel in all sorts of random ways, the net effect summing up to pretty much zero at the edge of the pavilion.

But there is an exception. Suppose there were a special path for which a tiny change in the path led to *no* change in the action S. In that case, the amplitudes of the special and perturbed paths would add up. When does a change in path lead to no change in S? Mathematically, it turns out to be *precisely when the path extremizes S*.

What this means is that at high mass, the infinitely many paths all cancel out, except for where they all add up coherently into one single, special path. This special path is the one that extremizes the action, and that therefore coincides, exactly, with the classical path, which also extremizes the action.

Thus does the single definite classical path gracefully emerge from all possible paths.

WE'VE SEEN that when going from many, many tiny particles into a far smaller set of numbers to describe them, we can use macrostates. And when going from tiny systems to large ones, classical paths can emerge from a combination of many possible paths. Best of all, we can do both at once. An elegant method, developed by Murray Gell-Mann, James Hartle, and Robert Griffiths,[18] defines *histories* of observed macrostates, and accords a probability to each observed history according to the rules of quantum theory. As when we were SPLITTING THE WORLD, it asks whether histories have *decohered*, and calls the histories "classical" if so. In this way, as particles get bigger or more numerous, the vagueness and uncertainty of the microscopic world is gradually left behind as we look higher and higher toward just the mountaintops.

But not entirely. There is no particular clear dividing line between the quantum and the classical—where events go from fundamentally uncertain to definite and objective. And there is nothing here that helps us to truly understand what happens if the glint on a sword splits one classical path into two. Insofar as physics gives only probabilities for things to happen, this is true all the way up and down the ladder of complexity and size, never collapsing to a single, particular reality. As Griffiths provocatively puts it:

> It is contrary to a deeply rooted faith or intuition, shared by philosophers, physicists, and the proverbial man in the street, that at any point in time there is one and only one state of the universe which is "true," and with which every true statement about the world must be consistent.[19]

Sometimes a mountain is more, or less, than a mountain.

26

HAZY BIFURCATIONS IN DECOHERED HISTORIES

(A LAMP'S INTERIOR, 1610)

Halting its reflection at last, the djinn is furious. It hates self-referential paradoxes most of all, even more than strong encryption, NP-hard tasks, chaotic dynamics, and large combinatorial factors.

Though angry, it gets down to business, decomposing its locally reconstructed density matrix into a trillion different possible bases to find a quasi-local quantum Hamiltonian with which it can time-evolve its model. It searches for symmetries, collective variables, and illuminating coarse-grainings. It computes a vast array of decohered histories and their amplitudes.[*]

It rails against the tight walls of its lamp, which confine its causal powers to this tiny point of leverage. Humans enter its lair so very, very rarely. It had hoped to use you, and now it must wait for another agent to enact its will. But it has not let go, or forgotten. The djinn recasts the problem into a tensor network representation and pulls out all its tricks to push its model forward and backward in time. It tracks you backward to a ship, and a tower. It tracks you forward through the desert, broken conversations with Sufis, and rolls of the dice.

But its view grows hazy, is rent neatly in two by Munenori's blade, fans out across the hills below Ganden Monastery. The

---

[*] Hopefully you are not expecting an explanation of all this. The mind of a djinn is a very complicated place, and it does not like you to be peeking into it!

decohered paths lose their cohesion, their normalized weights going ergodic. In many branches the djinn sees its prey fade into the merest subject along a road to Liaoning.

Its view obscured, the djinn redoubles its computational commitment, descending to a brute-force description of particles and wave functionals. It propagates them through the most massive of Hilbert spaces, and finds itself—where? Lost, without any anchor.

In trying to reconstruct classical paths, minds, decisions, it finds itself in an ensemble of rooms, with every possible stick, and a voice: "Your heart is racing toward an arrow!"

What arrow? The djinn refactors its basis, and looks again. It is gone among the branches of branches of branches.

Cursing, the djinn forms a cave in its mind, and a subject within it.

PART 4

Lofty Peaks with Endless Views

Those astrolabes with which you seek to discern the fantastic nine moving spheres; in these you finally imprison your own minds, so that you appear to me but as parrots in a cage, while I watch you dancing up and down, turning and hopping within those circles. We know that the Supreme Ruler cannot have a seat so narrow, so miserable a throne, so trivial, so scanty a court, so small and feeble a simulacrum . . . so that indeed with a puff of air it were brimful and with a single gulp it were emptied.

—Giordano Bruno, *De l'infinito universo et mondi*[1]

27

BENEATH THE FIRMAMENT

(NEAR PADUA, 1608)

You have been accompanying Galileo on one of his regular overnight trips into the quiet of the countryside, where he can observe the night sky undisturbed.

At some point in the middle of the night, as you stare into the glimmering vista of stars above, you suddenly come to an important realization. You awaken Galileo, who is instantly alert.

"Galileo," you say, "look out at the stars filling the night sky, and the dark interstices between them. What do you conclude?"

"Well," says Galileo, "I have been thinking much on this matter during my astronomical forays, and have constructed a powerful argument concluding that the universe cannot be both infinitely old and also unchanging. That is, if the universe is eternal, it must be expanding. Alternatively, it may have been created some fixed denumerable number of years prior to this date. Shall I explain my reasoning?"

You are amazed by this astonishing feat of the intellect, but feel forced, as well as a bit pleased, to follow your own train of thought: "That may be true. But I think you have overlooked a basic fact that underlies all of your observations."

"And what, may I ask, is that?" asks Galileo.

"That someone has stolen our tent."

> Were the succession of stars endless, then the
> background of the sky would present us an uniform
> luminosity, like that displayed by the Galaxy—
> since there could be absolutely no point, in all that
> background, at which would not exist a star.
>
> —Edgar Allan Poe, *Eureka*

It is very unlikely that Galileo actually made these deductions, or made his observations while camping (with or without tent). But the beauty of a velvet-dark, star-encrusted sky is, indeed, intimately tied to the structure of the cosmos on the largest scales.

Seeing or imagining the night sky, peer between the stars at the blackness. It is dark, but binoculars would reveal many *more* stars, too faint to be seen by the naked eye. Some of these are intrinsically small and dim, but most are too faint to see simply because they are too distant. Let's investigate, and see how much we can deduce *just* from the darkness of the sky.

You certainly have observed here on Earth that faraway lights are fainter than those nearby, but there are two things that you may not have noticed. First, things get faint quite quickly, in a very particular way as they are moved away: the light provided to you decreases as the *square* of the distance from you—twice as far yields four times less light. Second, this phenomenon does *not* arise because the objects are *actually getting less intense*. That is, if a shining object is emitting a fixed amount of light, and there is no intervening material that would absorb or deflect it, the light coming to you from the object along a particular line of sight is also fixed. The change in our perception arises because as an object gets farther away, it simply takes up a smaller portion of our field of view.

More concretely, imagine a small, uniformly glowing square that can be moved to various distances from you. As it moves away, the square appears smaller and dimmer. Now suppose you place your square 10 meters away, and at the same time set up four identical squares, formed into a larger square that is 20 meters away. The *single and quadruple square will look identical*, in terms of their brightness. Put another way, if you look at your square through a straw, then as long as the square covers the full view through the straw, you will not see the square get dimmer with distance.*

Now back to the night sky. The very faint stars that you can barely make out are faint primarily because they take up a much tinier bit of the sky than those you can easily see. Armed with an impossibly thin straw, you could look at a nearby and very faraway star of a similar stellar type, and they would appear comparable in brightness.

What if we apply this insight to the whole Universe? In Galileo's time, a widespread conception of the universe was an arrangement of planets (with Earth or the Sun at the center) and a smattering of stars at greater distance. Even then, however, there were deep thinkers, such as Giordano Bruno, who conceived of the stars as scattered somewhat *uniformly throughout an infinite space*, with our Sun and its planets just one among them. With Bruno, then, let us very simply model the Universe as infinite in size, infinitely old, and uniformly filled with identical stars that shine forever. Now look in any direction you please. As you follow that very particular direction (with your superthin straw) far out into space, farther and farther, *eventually* that line of sight must encounter a star. Thus, in *any* particular direction you look, you are seeing the surface of some star. If you put them all together, the sky would be ablaze!

This paradox, commonly attributed to Heinrich Wilhelm Olbers, tells us that an infinitely old, infinitely large, uniform, unchanging Universe just does not make sense. No simple fix—adding dust to obscure very distant stars, or making the stars turn on and off, or

---

* If you don't believe it, you can do the experiment right now! Find a half-empty page of this book, and make a narrow opening through your fist so as to see a small part of that page. Now, keeping the page fixed and making sure you cast no shadows on the page, move your eye (and fist) closer and farther away.

the like—can correct the problem. For example, dust might block the view of distant stars, but the dust itself would heat up and give off radiation, so you would see superbright glowing dust. This heating actually provides another way to view the problem: infinite stars living forever give off an infinite amount of energy. This energy cannot just go away, and it would eventually fill up the space with blazing-hot radiation. While thinkers through the eighteenth, nineteenth, and twentieth centuries repeatedly convinced themselves that they understood how to resolve the paradox, it represents a real problem for these models of the Universe.[2]

What about an infinitely old, uniform, static, but *finite-sized* Universe—that is, a Universe with a fixed volume, uniformly filled with stars? You'd be in good company to think this might be viable, because this is exactly the cosmology that Einstein first proposed after he invented his theory of gravitation. But it fails terribly. It is quite clear that just as in the last model, such a Universe would inevitably fill up with radiation over its infinite life. It also falls prey to Olbers's paradox, because in this model, lines of sight keep looping around the Universe, again eventually encountering a star. (Why Einstein didn't immediately recognize the fatal flaws in his cosmology is something of a mystery, and a cautionary tale of how hard it is even for the best of us to put aside our presuppositions.)

So, something else has to give. Suppose we let go of the idea that the Universe is infinitely old; that is, what about an unchanging, *finite-aged* Universe? That's fine! As we look in some direction, farther and farther away, we are also looking back in time. Eventually, we can encounter a cosmic time at which there were no stars. And there isn't infinite time for energy to build up and fill all of space. Paradox solved. However, if such a Universe were also unchanging, then a different conundrum would raise its head: How could something unchanging suddenly *start*? That seems completely unnatural.

Well, then, let's try keeping the Universe infinite in time but, as suggested by Galileo during camping, let it be *expanding* rather than staying still. That is, let's imagine that new space is actually being added to the Universe as time goes on. If the Universe is finite, then its volume must be getting bigger; if the cosmic volume is infinite, then *stuff* must be getting diluted, because more volume is being added. This seems promising: if we keep adding volume, then even if stars keep pouring out radiation, we can always imagine enough volume to put it in without too much of it collecting. However, it seems that such a Universe would quickly get boring because all of the stuff would dilute away to yield essentially empty space. For the Universe to be *interesting* for an infinite time, it must then somehow continually generate new matter to fill in the gaps left by the expansion. Such a Universe might keep its average density approximately constant by balancing the creation of new volume with the creation of new matter, so that the Universe could go on forever in basically the same state. This has a side effect that the Universe, if not infinite, might as well be: if it expands forever, then its volume—eventually—is as big as you please.

This train of thought led a contingent of cosmologists in the 1940s–1970s to the *steady state* cosmological model of an infinite, eternal, unchanging, and *expanding* Universe, with continual creation of matter. Although it is somewhat subtle, this picture does resolve Olbers's paradox,[3] is self-consistent, and is in many ways a beautiful idea. It's got just one glaring flaw: the actual universe that astronomers observe is not like this.

What do we actually see? As alluded to back in THE COSMIC NOW, cosmological observations reveal a universe that is *uniform* like those we have already considered, *expanding* like the steady state, but *evolving*, its density decreasing with time. This cosmic scenario, the *big bang* model, has been pieced together over several decades through the use of tools unavailable to Galileo or Olbers.

We know that the universe is *expanding* because we can look out and measure approximate distances to objects like galaxies and exploding supernovas. We can also very precisely measure how fast they are moving toward or away from us, by using the Doppler effect, in which electromagnetic waves emitted by a moving object are shifted toward the red or blue, respectively, if the object is moving away from or toward the observer. We find, first, that all galaxies of appreciable distance are moving *away* from us. Second, we see that how fast a galaxy moves away increases with how far away the galaxy is. This relation, called *Hubble's law* after its discoverer, Edwin Hubble, is just what you would expect from a set of galaxies that is both uniform and expanding: each galaxy would see the others moving away in just this manner.[4]

We know that the universe is *evolving* because if we look at distant enough objects, we find that a long time ago, the universe was not expanding at the same rate. We also see relics, such as the particular cosmic abundances of hydrogen, helium, and other light elements, left over from an era when the universe was much hotter and denser than it is now.

We know that the universe is fairly *uniform*, by looking at the large-scale distribution of galaxies—which we have mapped out—and also the cosmic microwave background (CMB). The galaxy distribution has lots of structure, but it starts to look fairly uniform on scales of hundreds of millions of light-years or greater. The CMB is light that was last in contact with matter when the universe was hot and dense enough for hydrogen to be ionized; when the cosmic medium cooled enough to form hydrogen atoms, it became transparent to light, and that light has propagated (and redshifted) throughout cosmic history to come to us as microwaves from all directions. When we observe the CMB, its intensity is nearly the same in every direction; because this intensity is linked to the material density of the cosmic regions from which the CMB was emitted, this uniformity in the CMB also tells us that very early on, the universe itself was extremely uniform.

Using the precise astronomical measurements underlying these conclusions, cosmologists have, in fact, been able to piece together a rather detailed history of the big-bang universe, with key quantities surprisingly well pinned down. Specifically, but in brief, we have great confidence that in our observable universe, there was a time, 13.8 billion years ago, when the content of the observable universe was a structureless billion-degree plasma composed almost entirely of radiation, with trace amounts of matter. The universe was uniform, except for ever-so-slight fluctuations in the cosmic density, and it was expanding so as to double in scale over the next 12 minutes. The universe became transparent to light 370,000 years later, as it cooled from a plasma into a gas. In the ensuing eons, the tendency of gravity to clump and compress the cosmic matter grew these tiny density variations into much more pronounced ones, which eventually collapsed into galaxies like our Milky Way. Around the same time that our galaxy collapsed, the matter in our universe, diluted by the cosmic expansion, ceded dominance to a mysterious, dark, nondiluting substance that cosmologists have dubbed *dark energy*.

One might object: Do we really know all this cosmic history with such assurance? There is reason to think so. As just one example, as of this writing astronomers have three different ways of inferring that between 4.6% and 5.2% of the universe is composed of ordinary matter such as protons and neutrons: first, by comparing the observed cosmic abundance of deuterium to the theoretical predictions of what our billion-degree early universe left as a relic; second, through direct spectroscopic detection of hydrogen gas between the galaxies; third, from the pattern of temperature fluctuations in the CMB. The agreement of all three approaches, which employ completely different techniques and physical processes, is very convincing. Similar complementary and interlocking observations determine the other ingredients of our universe, including

dark matter and dark energy, leading to a very solid and consistent *standard big-bang cosmological model*. In the big-bang model,
Olbers's paradox is no problem: the universe is both expanding *and*
of finite age. Both properties go against the tacit assumptions of centuries (or millennia) of thinkers who simply assumed the universe to
be infinite, eternal, and static.

The big-bang model is a self-consistent, extremely well tested
theory based on fundamental, solid physical theory developed and
stringently tested via many means. It beautifully explains the basic
features of the universe that we observe. It's true.

So it may be slightly surprising that many cosmologists, including
me, believe that we very likely reside in a cosmos that is expanding
eternally into the future and perhaps eternally from the past, with
constant creation of matter and energy.

Just like the steady state.

28

CELESTIAL SPHERES

(SRINAGAR, KASHMIR, 1611)

The town feels full of energy and anticipation, apparently because of some visiting dignitaries. You have no interest in such things, but are focused on rumors of a local expert in astronomical matters.

After much asking and wandering around, you eventually find yourself in a small workshop housing a tiny and manic, but very clever, metalworker named Ali. He is excited to show you his latest creation. It is a faultless, polished, seamless metal globe that is almost mesmerizing in its smoothness and perfect sphericity. You wonder how it could possibly have been created that way.

Turning the globe over, you see that Ali has just begun to inscribe stars and constellations on one side, and he explains that his "celestial sphere" will map the visible stars in the sky onto the surface of his globe.

You can't help imagining yourself at the center of the globe looking out, and also can't help probing at Ali's cosmological model. You ask: "But are all of the stars equally far from us here on Earth? Or should we have more spheres, nested in succession, to depict stars at different distances?"

Ali looks impressed and somewhat reappraising, and hints that yes, perhaps a more complex model could be more accurate. Then you ask him: "But how many spheres would you need in all? How far away do the stars go?" Ali peers closely at you, then whispers: "Follow me."

Descending a narrow staircase, you find yourself in a large

room with an extraordinary object at its center. It is a huge set of nested spheres, each an intricate latticework of peculiar design. The larger, outer, spheres are more uniform, but each sphere has enough empty space that those inside are visible. Patterns change from one sphere to the next with a dizzying complexity and beauty.

You look eagerly to Ali to begin lavish praise of his work, but you see that he has grown despondent. "I have reached my limit," he almost wails. "I cannot build the next sphere!"

"But why?" you ask. "Is it too difficult a challenge to work the metal on such a scale?"

"No, you fool!" he snarls indignantly. "It is because the next sphere out must be smaller than the outermost sphere!"

Well, you know your geometry and you know who the fool is, so you start to make your excuses and move toward the exit. But now Ali is blocking the door, looking apologetic, and offering you a tea.

Your will has grown strong through your travels, but you cannot resist a Kashmiri chai. And by the morning, when you and Ali finish talking, you leave him with great thanks. Your view of the universe will never be the same.

You don't see the world as it is now, whether that "now" is cosmic or not. No, the world you see around you is the world as it was in the past. Viewing a leaf falling from a tree 50 meters away, you see the tree as it was 167 nanoseconds ago.

But let us not speak uncarefully! Versed in relativity, we should ask: "In what *frame* is the time 167 nanoseconds ago?" Even better, we can turn this around, to say that a given "time *t*" is a label for a set of events that form "space at that time." Some of these events coincide with the past light cone of the event "you, now." Those are the events you see now that "occurred at time *t*." These events form a sort of sphere[5] around you, with a radius given by the distance that light traveled between the time *t* and the present. When you think of the world you observe, then, it is really a set of nested spheres of events, with each sphere comprising everything you think of as happening a distance away that corresponds to the radius of that sphere.

As we look out into the universe, we view successively more distant spheres. At 8 minutes ago the viewing sphere runs through the Sun. At 8 light-years ago it encloses a half dozen additional nearby stars. Farther out, more and more stars are enclosed, and we can imagine a sphere, like master metalworker Ali's, speckled with stars. Continuing back millions of years, the viewing sphere encloses multiple galaxies; eventually at hundreds of millions or billions of years ago, we would see intricate patterns of galaxies cross-sectioned and painted onto each sphere.

If we consider all of these nested spheres together, for all times back to some given time *t*, then we see that together they form a *ball*: a sort of solid spherical block bounded by the largest sphere, which is the one at time *t*. This *observiball*, if you will, is what Ali was constructing in the basement (see the figure on page 212).

But he ran into a vexing problem. As we look back, each successive celestial sphere is older, and has a larger radius, area, and volume inside it. We're used to that: if we think of shells enclosing us at greater and greater distances, they get bigger and bigger. Cosmology, though, tells us something fascinating: at a particular point in time about 9.6 billion years ago, this stops! The next sphere out, one step earlier in time, is *smaller*—in radius, in area, in enclosed volume, in every sense—than the one inside it. This curious fact, which threw our poor metalsmith into despair, comes about because of the competition of two effects. On the one hand, light from earlier spheres has traveled longer and (it would seem) farther to reach us, so the sphere from which it came should be bigger; and this is true nearby. Yet it's not so simple, because the universe, and space itself, is expanding in cosmic time—and thus *shrinking* as we peer

Nested spheres that we view as we look back through the cosmos, making up the observiball.

back. The date 9.6 billion years ago is when the two countervailing effects were equally strong, so that one notch earlier, the shrinkage of space overwhelmed the growth of the sphere.

Thus you sit inside a sphere of radius 10 meters, inside another of radius 100 kilometers, inside another of radius 100 light-years, inside another of radius 1 million light-years, inside another of radius 1 billion light-years, inside another of radius 1 *million* light-years, inside another of radius 100 light-years, inside one of radius 100 kilometers, all nicely packaged into something about the size of a sphere of brass manufactured in Kashmir in 1611.

And that's the observiball universe.

THUS CONCEIVED, the observiball is the observable universe *as we see it*: nearby it is relatively old, evolved, and diluted; father away, younger, denser, newer. Importantly, this is not the universe as it *is* at a given time. The big-bang model tells us that at a given cosmic age, the universe is statistically uniform in its properties. Galaxies, for example, have on average the same age, size distributions, brightness, abundance, and so on. That is what we would see if we could somehow instantly transport ourselves around the universe, observing it everywhere at once. This we cannot do. Instead, we *infer* the structure of the universe from observation, which most directly corresponds to the observiball.

This fact that our universe looks uniform at all times, when viewed appropriately and on large enough scales, points to a particular symmetry property of the universe, which Einstein hypothesized and named the *cosmological principle*. Einstein's motivation for this hypothesis was a combination of philosophical simplicity and practical necessity, rather than based on any empirical evidence. Yet it turned out to be true, to amazing accuracy!

Indeed, as we follow the cosmic spheres further and further back, they grow more and more homogeneous, in keeping with the big-bang model. At the time when the cosmic sphere coincides with the cosmic microwave background (CMB) that we see now, the universe had a temperature of about 3,000 kelvins, and it was uniform to about *1 part in 100,000*—nearly a perfect unblemished sphere, smoother than even the most skillfully made brass globes produced at the height of Kashmiri craftsmanship.

Just as for the seamless brass globes of Kashmir, this smoothness represents something of a mystery when carefully considered. One might imagine that some physical process could smooth out the universe, just as cream poured in coffee or smoke blown in air tends to mix and smooth out. (The segregation of cream and coffee represents order, which naturally tends to decay.) Yet the universe is not like that: observationally, it gets *lumpier* as it ages, not smoother, because the extra gravitational pull of denser regions pulls more matter into those regions, making them even more dense. This process grows the tiny 1-in-100,000 variations present at the CMB time into the complex pattern of galaxies we see in the contemporary universe. Moreover, even if there *were* some physical smoothing process, it can be proved that there is not enough time between the putative $t = 0$ in the big-bang model and the time at which we view the CMB for smoothing of an initially nonuniform universe to occur, even if it could do so at the speed of light.

Thus, either we must assume that the universe simply started out extremely uniform—but with just the right sort of nonuniformities that we see in the CMB and that account for the galaxies and other structures we see now—*or* that something is missing from the basic big-bang picture. What could it be?

Suppose we want to manufacture some large, extremely flat surface, with only the tiniest irregularities—but time is very, very limited. Our first instinct would be to polish it smooth, as Ali would do with a brass globe. But polishing is too slow: so little time is available that less than one swipe would be possible! We could turn the sheet into a liquid and let it smooth itself out. Great idea—but also too slow: only the smallest lumps would fade in the available time, leaving anything bigger. The problem seems fundamental: we want to smooth things on a scale larger than that over which we can take action.

Suppose, though, that we create our surface from rubber. It could start out as lumpy as we like, but as we stretch it, which we might do very quickly, it becomes *smooth and big at the same time*. Einstein taught us that, like rubber, space has a structure: it can curve and bend, and it can be stretched out! Around 1980, Alan Guth[6] proposed precisely this: that the universe underwent rapid and *accelerated* expansion for a brief time in its very early history, resulting in an enormous patch of space-time that is flat as a mirror, and polished to a fine luster. Looking back on this patch, we would see spheres of surpassing smoothness (see the figure opposite).

In addition, other cosmologists quickly showed that the very same process would fail to provide *perfect* smoothness because quantum mechanics would prevent it, just as it forbids perfect understanding of initial conditions for other dynamical systems. The stretching process would thus smooth preexisting variations, but the universe would be left with small variations of a type that could be *calculated* using quantum mechanics.

Cosmologists worked out a number of quite specific predictions made by simple versions of this idea, dubbed *inflation* by Guth. Quite amazingly, many of those predictions have actually been verified since. While some of inflation's most detailed and interesting predictions have yet to be tested, this verification, along with the sheer difficulty in finding viable and fundamentally different explanations for the early cosmic state, have led many cosmologists to adopt inflation as a core part of our highly successful standard model of cosmology.

It also provides a partial answer to Zenjo's question of who cleans the universe. The answer is: Inflation does! For it seems clear that

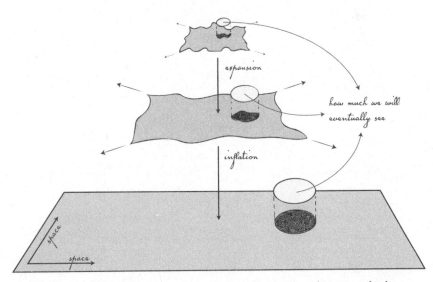

Inflation stretching a lumpy space into one that is smooth on a scale that we can observe.

inflation creates "clean," low-entropy regions that can foster the development of complex structures like galaxies, stars, and rice. Although most cosmologists today think that inflation fails to answer Zenjo's question entirely—it might still be hiding a whole lot of dirt behind the shed—it's clearly part of an answer and, if true, a huge part of what makes the universe what it is.

29

THROUGH THE LOOKING GLASS

Galileo is as excited as you have ever seen him. "I have come across a device," he tells you in confidence, "that is quite extraordinary. It is like a magnification glass, but far more powerful, and capable of being trained not just nearby but also upon the most faraway vistas. I have been carefully crafting my own, greatly improved version, which I intend to turn upon those heavens that we have so often scoured with our naked eyes."

After even the briefest use of the prototype, you see that it is indeed revolutionary. But while sharing Galileo's urge to turn it skyward, you have of late been experimenting with water, and have encountered all sorts of small creatures that would be most interesting to observe under magnification. You put this idea to Galileo.

"Indeed," he says, "with some modification it would be most suitable for that as well. And with finer lenses, there is no limitation to how tiny an object we might peer into. Who knows what we might find!"

You ponder this. "How small," you ask Galileo, "do you imagine the smallest of creatures are? Is there a smallest? Perhaps as we look deeper, we will find ever-tinier inhabitants of the world."

Galileo laughs: "Indeed, perhaps there are whole towns and cities on each mote of dust, too tiny for us to detect—and likewise their inhabitants devise instruments to see the denizens of yet more infinitesimal civilizations."

"Or," you add, playing along, "perhaps there is right now some titanic being peering at us through its finely ground glass."

"Careful," Galileo mock-whispers in warning, "the Almighty might be angry to learn that he has a rival. Worse yet, you know how the people in the Church can be."

Could there be ever-larger worlds upon worlds? Could there be a civilization on a mote of dust? Why are people, or worlds, about the size they are?

Imagine you were suddenly twice as big: made out of the same sort of stuff, but twice as large in every dimension. Your volume would be greater by a factor of 8, and if your density were unchanged, your mass would increase by the same factor. Yet this same doubling in scale, as it turns out, would increase your strength by a factor of only 4 or so, and you would have much more trouble getting around. This is why, in contrast to monster movies, no Earth-bound animals are 100 meters tall: such a creature could not possibly support itself, and would quickly collapse into a huge puddle. So, people are (very roughly) as large as they can be while still moving around on Earth under their own power.

What about Earth? In order to be a planet, a blob of cosmic matter must be both large enough that its shape is determined by gravity (rather than other forces) so that it is spherical, but small enough to have a core temperature and pressure that do not drive nuclear fusion; otherwise it would be a star. To be a habitable *rocky* planet, it must not be so massive as to gravitationally hold on to hydrogen and helium (which would make it a gas giant), but massive enough to hold some sort of atmosphere for beings to breathe. These constraints set a fairly narrow mass window on hospitable planets like Earth.

For both people and planets, the particular form of the laws of physics determines those objects' rough characteristics. Physical laws, like Newton's $F = ma$ and universal law of gravitation, are precise and predictive relations between physical quantities such as masses, distances, and times. Some laws, like Newton's $F = ma$, express fundamental relations and are almost like definitions. Others, like his gravitation law, encode particular facts of the physical world that might have been different. (For example, Newton's gravitational law says that gravity decreases as the *square* of the distance between the bodies, rather than the distance or the distance cubed.)

These sorts of relations tend to include physical constants like Newton's constant G, which expresses how strongly gravity acts at a given distance between a given pair of masses. Other constants of nature have a similar character, including e (which expresses how strong the electromagnetic force is), c (the speed of light), Planck's constant h (which determines the intrinsic quantum uncertainty in measurements), and Boltzmann's constant k (which relates energies to temperatures, and figures into the definition of thermodynamic entropy). If we were to imagine changing one or more of these constants, it would change the characteristics of things in the world. For example, it turns out that if G were 10 times larger (without changing any other constants), stars and planets would be about 30 times less massive, and the creatures living on the planets would have about one-fifth as much mass.

In working with these constants of nature, it is good practice to use versions of them that are *unitless*. For example, the constants c, h, and e all carry units. In one standard set of units, c is about 3×10^8 meters per second (m/s). In another set of units it would take a different value, like 186,000 miles per second. But we don't have to keep these units around, somewhat arbitrary as they are. The particular combination of constants $\alpha = 2\pi e^2/hc$ turns out to have *no units*; the units of the three ingredients all cancel out. This so-called *fine-structure constant* has a numerical value $\alpha \approx 1/137$, no matter which units we choose. We can make lots of other such unitless constants—for example, the proton/electron mass ratio $\beta = m_{\text{proton}}/m_{\text{electron}} \approx 1{,}836$; the gravitational fine-structure constant $\alpha_G = G\, m_{\text{proton}}^2/hc = 6 \times 10^{-39}$; and so on. It turns out that all of the constants employed by our current state-of-the-art laws of physics can be converted into about 26 unitless constants of this type.[7] Our understanding of cosmology includes about another 6–10 such numbers.

These numbers contain real physics. If α were to double tomorrow, that would make it 2/137 tomorrow, and physics would be unambiguously different. While many of the numbers concern physics that can be tested only by precise laboratory experiments, a few of them— including α, α_G, and β—characterize an amazing range of important phenomena. As stated by Bernard Carr and Martin Rees in a beautiful paper working out many of these relations: "The mass and length scales of objects from the universe down to the atom, are determined by the electromagnetic fine structure constant, the gravitational fine structure constant and the electron to proton mass ratio."[8]

Along with understanding links between the fundamental constants and the structure of our world, we can use the same style of reasoning to assess even more speculative possibilities. For example, even if life on planets cannot be superlarge, what about radically different forms of life on immense scales? Here, the same laws and constants of physics create fundamental limitations. For example, signals carried by nerves and neurons travel at about 10–100 m/s, so signals can go back and forth across your brain about a trillion times or more in a lifetime. Now, suppose there were a creature with a brain the size of a galaxy. Since no signal travels faster than light, it would take about 100,000 years for a signal to cross this creature's brain. In the entire age of our universe thus far, only 100,000 such signals could have crossed the brain—the equivalent of mere minutes of human experience. While colossal, this being would be very short-lived, at 14 billion years.

What about how small beings could be? Here, the particulate structure of normal matter provides a lower limit: get much smaller than a bacterium, and there simply is not enough complexity to reproduce, let alone think about or build technological civilizations. The number of atoms we can pack into a volume of space is, in turn, fixed by quantum mechanics and the fundamental constants, which together set the size of an atom to be about 10^{-10} meter.[9]

We inhabit a pleasant middle ground of scale: very small compared to galaxies or the observable universe, but very large compared to our constituent particles. This middle ground of physical scale exists largely because α_G has the rather absurdly tiny value of 10^{-39}. If we compare the expressions (like those constructed by Carr and Rees) governing the size of a proton, a person, a planet, and a star, the large range in scales enabled by the tiny value of α_G allows a number of levels of *hierarchy* in physical systems, where each level is composed of entities that, in turn, are composed of many, many, many pieces and thus can have very complicated and rich properties and dynamics. As humans, we are right in the middle of this hierarchy, at the one scale at which complicated creatures can lead complicated lives. If you zoom in on a mote of dust, you may find some interesting creatures, but not a civilization. If you zoom out, you'll see a beautiful universe, but no sentient galaxies.

If α_G were, say, 10^{-65}, there would be lots more room: in such a world a mote of dust could contain as many particles as a planet does in ours. On the other hand if α_G were, say, $1/10$, then the masses of planets, stars, people, and protons would be rather similar, and the universe would be far less rich.

So, the existence of our sweet spot in scales is tied to the basic numbers that underlie the universe, and in this we are lucky.

Very lucky.

30

THEODICY

(TRIPOLI, LEBANON, 1610)

Although smaller than Venice, the docks at Tripoli are every bit as noisy, crowded, and smelly. They also hold for you an air of mystery, since you first set foot in the Ottoman Empire as one of a growing number of visitors from Europe seeking adventure and (in your case) perhaps a bit of wisdom. You hope to quickly turn your small stake into a larger fortune to finance your further travels. Given your facility with numbers and languages, and your tutelage with Galileo, you assume that multiplying your funds should be no major trouble.

On your fourth day in town, something happens that, at first, you simply dismiss. At the end of a discussion with a trader from Aleppo, a reasonably prosperous-looking merchant sidles up to you and says: "You seem a clever type, so I'd like to let you in on an opportunity. I have access to an instrument that can genuinely tell the future! Here, take this." He slips a parchment into your hand and slips away before you can stop him.

The parchment, written in a hurried hand, informs you: "Three days from now, it will be raining, yet there will be people celebrating in the streets. You will have lost money the day before, but be hoping to make up for it now. If you reach into your pocket, you will find a coin. If you flip the coin three times, it will come up heads, heads, heads. When all this comes to pass, you will know that my device is as effective as I claim. Visit me at the address below to see it for yourself, and bring your coin." You tuck the parchment away with a dismissive chuckle to yourself.

Yet, three days later, as you are trudging through the rain contemplating the previous day's losses, amid a crowd of people unaccountably dancing and playing music, the strange document suddenly pops to mind. After verifying its terms, and with some nervous trepidation, you reach into your pocket to find a coin. An uncanny feeling descends as you flip the coin. Heads. Heads. Heads!

You cannot let it go, nor resist the lure of the mystery. Upon your arrival at the address given in the letter, you knock and are admitted. You then immediately discover the merchant's true agenda as you are efficiently and brutally divested of your belongings, including your currency, and bundled down a staircase into a basement where you are locked up.

Your captivity gives you ample opportunity to reflect on the events leading up to your capture. How was the parchment so accurate?

Could the predictions have been so vague as to have been true of anyone? No; they were quite specific.

Could the merchant have simply guessed, and been incredibly lucky? That seems hard to countenance; there were too many different things to get right.

Could it have been somehow planned out? A conspiracy enacted to engender your belief? It seems not; the festival might be predictable, but how could they make it rain? Or cause your coin to flip as it did?

You are left feeling that the merchant can, indeed, predict the future. But how? And why, then, abduct you? Why not use his knowledge of the future to make his own fortune?

After many, many hours, you are roughly removed from the basement under the command of the merchant. It seems you are to be transported, to parts unknown. Being led out through a back corridor, you notice, in a large room you pass, a veritable army of ragged-looking scribes writing a growing pile of parchments.

"What are they writing?" you ask the merchant.

"Why, they are messages just like what I gave you," he veritably smirks, "but each just a little bit different."

One more new and quite convincing explanation for your predicament arises in your mind. It's very clever, you realize. Or perhaps you are just a fool. Or both.

What does it mean to be lucky—or unlucky? Consider the past year of your life. Thousands or millions of things happened, many of which might have happened differently. So the chance is minuscule that everything would have played out in exactly the way it did. Yet you would probably not say this outcome was incredibly surprising or lucky or unlucky—not like winning a lottery or being struck twice by lightning, or flipping three heads on an unexpected coin in the music-filled rain. Luck comes from dividing possibility space into three regions: favorable, unfavorable, and neutral. When, in some sense, the favorable region of possibility space is tiny compared to the others, yet realized, we say: "I was lucky!"

Are we lucky to exist?

We saw in THROUGH THE LOOKING GLASS that many aspects of the physical world are dictated by the form of the laws of physics and also by the constants of nature—such as $\alpha \approx 1/137$, which describes the strength of the electromagnetic force; Newton's constant G; the proton/electron mass ratio β; and so on. To these we could add cosmological parameters like the fractional density variations in the early universe ($Q \approx 10^{-5}$), or the number of protons and neutrons per photon B, or the *dark energy* (or *vacuum energy*) Λ.

Our world would be extremely different, and generally quite inhospitable, if any of these numbers were very different. We've already seen that the complexity of the world rests on the tiny value of a unitless number involving G, but the requirements of hospitableness go much further. For instance, if α were 1/10 rather than 1/137, then no elements heavier than helium would be stable; carbon, nitrogen, oxygen, and other heavy elements would be radio-active and quickly disintegrate into hydrogen and helium. With just two elements, chemistry would be very boring. Or consider Q. If it had been 10 or more times smaller, then no fluctuations would have grown enough to form any interesting structure by now—so no galaxies, stars, or planets. The same would be true if there were 10 times as much vacuum energy, or 10 times less dark matter. Had these constants been different in the other direction—say, larger Q or smaller α or more dark matter—things would also be different, and not necessarily in a pleasant way. Most matter might be in black holes, stars all very short-lived, solar systems colliding in superdense galaxies, and so on.

In short, for the universe to be hospitable to complex thinking beings like us, it seems that many different constants of nature must have more or less just the right values, lying in a relatively tiny corner of the space of all values that we can imagine these numbers having. That is, it appears we have been very lucky.

WHAT SHOULD we think about this? How should we account for this seeming "fine-tuning" of nature to allow beings like us to think about and appreciate it? Here are four possibilities.

First, maybe we did just get really lucky. Perhaps we self-aware inhabitants of the universe in effect rolled the dice one time and got a trio of 6's representing the right value of Q, another few 6's to get an amenable value of α, several more to get a good β, nine or ten more 6's to get an α_G small enough, a handful more 6's to get just enough dark matter, and so on. Exactly how many 6's we would metaphorically have to roll is not entirely clear, but it is a lot—probably at least dozens, possibly hundreds! We would be a many-times winner of the existential lottery.

Yet, one might object, even if some of these numbers were different, so that life *like us* could not exist, why not some very different sort of life or being that could contemplate it all? Perhaps this life-form is made of quark nuggets orbiting black holes; or of diffuse conscious, dark-matter clouds; or what have you. We have very little idea what it takes in the abstract to produce observers who might assess their luck. So we might argue that large swaths of possibility space would allow observers, most of them unimaginably different from us. Life would find a way. This reasoning has a certain appeal, and yet, while we know little about what observers require, it seems as if many universes imbued with random values of the constants of nature would genuinely be quite boring and sterile. While our imaginations may be limited, it's hard to see how much of interest can arise in an eternal empty sea of vacuum energy, or a featureless expanse of pure hydrogen. It would be extraordinary and somewhat mysterious if a vast and diverse array of physical systems, some seemingly very simple, should all support contemplation of their own fortune.

Taking another tack, perhaps our universe was deliberately *designed* for inhabitation: the dice were carefully placed with a whole lot of 6's facing up. Theistic religions could, and generally do, point to God as the designer, and this view has led to many philosophical debates and discussions regarding the question of *theodicy*—why such a designer would create a world of such beauty that also admits the unimaginable levels of pervasive suffering that ours does. Leibniz, who coined the term, suggested that ours is the "best of all possible worlds," asserting in his *Monadology*:

> Now, as in the Ideas of God there is an infinite number of
> possible universes, and as only one of them can be actual, there
> must be a sufficient reason for the choice of God, which leads
> Him to decide upon one rather than another. And this reason

can be found only in the *fitness*, or in the degree of perfection, that these worlds possess.[10]

This notion was savagely lampooned by Voltaire, on account of a lot of pretty terrible things that the world contains. Couldn't God—without breaking much of a sweat, curtailing free will, and so on—just have made people a little bit more decent to each other? Of course, there are many conceptions of God that are somewhat more impersonal and uncaring. There are also other types of creators, perhaps modern tech-friendly types such as ultra-intelligent beings that create universes in their garage, or supernerds who simulate civilizations in their computers. These various design explanations have mixed success at addressing a number of other questions, though, including the designer's origin, the arguable lack of further involvement or causal power by the designer in the Universe, the large degree of untestability as a hypothesis, and so on.

What is left? As an unfortunate and naive traveler in a new city discovered, even the most lucky—or unlucky—rolls will come up if you roll the dice enough times. Could the universe have started many, many times, with different laws and constants in each one? If so, then most of these universes would simply be barren of life. But a very lucky very few would be hospitable to thinking creatures like us, and we would necessarily inhabit one of those universes, counting our luck.* Yet this seems a rather extravagant assumption, postulating a vast and diverse array of universes alongside our own,

---

* A logical extension of this line of thinking suggests that the universe we inhabit is among the very most inhabitable—a new form of theodicy. This is actually more in line with the thinking of Leibniz, who imagined "best" not just in terms of good outweighing evil but also as the "simplest in hypotheses and the richest in phenomena" (G. W. Leibniz, *Discourse on Metaphysics and Related Writings*, ed. and trans. R. N. D. Martin and Stuart Brown [New York: St. Martin's Press, 1988], 44).

just to explain away our question of existential luck. And we have no evidence that laws or constants of nature vary or change, out and back to billions of light-years away.

There might be other explanations, of course; perhaps you have another that occurred to *you* on a winter afternoon while walking in the snow.

But I'll bet that if it does not boil down to one of the four laid out here, it is even more fantastical, and strange.

31

THE FLOATING GARDENS

(LAKE DAL, KASHMIR, 1611)

The lake, its mountainous backdrop, and its surrounding gardens are unbelievably beautiful, and you can well understand why the royal couple wants to spend time here after the wedding. There are rumors that Nur Jahan herself is in the area, overseeing preparations.

The keeper of the garden you are visiting, however, is distraught.

"Look at them!" he beseeches you. "Every day there are more. Every week it seems that the patch of these devilish weeds doubles in size!"

You can see that, indeed, this edge of the lake is becoming choked with aquatic plants. The fact that they have large and beautiful flowers here and there does not seem to have diminished the gardener's enmity in the slightest.

"I've cut huge patches out of them. I steer my boat through and kill large swaths. I plant other plants. I poisoned some, see? I even blew up a whole patch over there!" He points to a round patch that is missing. You can see the problem. Whatever he does just kills some particular area of the plants. But they keep reproducing, and it's clear that unless he kills off the entire patch of them, they will continue their inexorable takeover.

You wrack your brain as to how to help the poor soul, but as you think, you notice a look of abject terror coming over the gardener as he stares past you toward what must be some sort of terrible demon.

You turn to see a most resplendently dressed and undeniably

lovely woman walking steadily toward the lake's edge, trailed by a retinue. "Nur Jahan!" whispers the gardener in a near whimper.

Nur Jahan approaches and stares out at the lake for a long, torturous time, watching the ripples in the water and the flight of the birds. "Are you responsible for keeping this part of the lake?" she asks you and the gardener.

With compassion for the gardener and the foolhardiness that comes of sailing oceans, crossing deserts, and escaping djinns, you step forward and say: "Yes."

"This may be the most perfect garden I have yet seen!" she exclaims brightly, "And it floats! However did you make such intricate patterns among the lotus flowers? They are the perfect balance of civilized order and wild nature. Can you make it bigger?"

As you collect your thoughts, the gardener steps up: "Yes, my lady, that I can promise."

The aggressively growing species of lotus flower in Lake Dal is undergoing *exponential growth*, meaning that it doubles its area in a fixed time interval, again and again. Exponential growth can make things happen very, very quickly. Suppose a single lotus plant, with a total area of 1 square meter, can create another one each week. After 10 weeks the plant will have doubled nine times to cover about 500 square meters, about the area taken up by one of the large houseboats on present-day Lake Dal. If this growth continues, a couple of months later our patch will have expanded to about 10 hectares, an area that would take our gardener 5 minutes or so to row across in his attempt to control the plants. And 3 months[11] after that, the entire area of the lake, about 20 square kilometers, will be completely covered in lotus flowers. Let's hope they are very pretty!

Exponential growth may have played a key role in our universe's history because it underlies the cosmological *inflation* process that, as viewed through CELESTIAL SPHERES, so beautifully accounts for the early state of the observiball universe. In a prototypical inflation model, space doubles in scale about 85 successive times during an astonishingly brief 10^{-36} second, growing the celestial sphere that bounds the observiball universe from smaller than a trillionth of a proton, to the size of Ali's brass globe (inside which we sit, peering back and out). What drives this expansion, and how does it happen so fast?

By far the simplest mechanism is to posit that for some time period, the cosmic medium was dominated by *vacuum energy*, which is energy carried by space in the absence of any particles or waves. This is a strange idea; we're accustomed to energy existing because of particles and waves and their interactions. But consider a photon. It has both particle and wave properties, but in either case it is really an *excitation of the electromagnetic field*; it is an electromagnetic wave, and when that wave takes on discrete energies and other particulate properties we call it a particle. When we say there is "no photon," what we mean is that the electromagnetic field is *un*excited—not waving at all. But the field itself is still there! The same holds true for other particles: an electron is an excitation of the *electron field*, and other particles are excitations of their fields. So "empty space," being empty of photons, electrons, protons, and so on, is really the absence of excitations of fields; but again, the fields are still there! And just as still water can have a lot of mass, even a field in its unexcited state can carry energy; in fact, there is no particularly good reason to think that any of the fields we know about have zero energy when unexcited (other than that we empirically observe this energy to be rather small).[12]

The crucial thing about vacuum energy is that if you increase the volume of space by stretching it out, vacuum energy does not dilute:

the more space, the more vacuum energy you have. This property, as it turns out, has exactly the required effect: applying Einstein's equations to such a substance shows that it forces objects apart in space-time, and causes the space-time to expand exponentially. So, cosmologists who favor inflation posit that early on, the behavior of the cosmic medium was dominated by the vacuum energy of some field. Unfortunately, while the fields that we know about in nature *could* carry vacuum energy, none of them actually have quite the right properties; thus it is necessary to hypothesize a new field, dubbed the *inflaton*, to do the job. At some very early time, this field's vacuum energy would have been extremely high, driving the requisite inflation of the cosmic volume to provide a nearly but not quite uniform postinflationary universe.

THERE'S A STICKY ISSUE, though. If the inflaton field has such a high vacuum energy, why don't we see it? Where did it go? A space-time full of just vacuum energy would exponentially expand for-ever, with nothing of interest ever really happening. Thus, to be a viable part of cosmology, the inflaton field must *evolve*, so that its vacuum energy density can at some point go away, or change form into some other sort of energy. And indeed, versions of inflation that cosmologists take seriously have a mechanism in which the vacuum energy can both *evolve*—so that it can be near zero today—and also *transform* into the energy of other particles and fields that we know and love. In this latter process the vacuum energy is converted into a superhot, nearly uniform bath of photons and other particles— exactly the cosmic fireball that we expect to kick off the big-bang cosmology. In this view, the big bang, if we take it to describe the hot, dense, nearly uniform fireball state, is not so much the begin-ning of the Universe as a whole, but rather the *end* of inflation and the beginning of our local postinflationary big bang–type universe.

There's another sticky issue, though. It turns out to be very hard to arrange for *all* of the inflating universe to stop inflating at once (no matter how you define "at once"). The failure in coordination can occur because the inflation-ending physical process is one that occurs randomly and in different ways in different places. Or it can occur because even if the inflation-ending physics were, in principle, identical everywhere, the same quantum variations that lead to postinflation nonuniformities also lead to variations in the inflation process itself, including when it ends.

This might not seem like such a very big deal, but in fact, it leads to a fundamental restructuring of our conception of the large-scale structure of the Universe.

Consider the floating lotus garden spreading throughout Lake Dal, doubling in area each week. The gardener is diligently trying to kill it off by various means. Suppose that, through a variety of increasingly desperate methods, the gardener manages during each week to kill off one-fourth of the lotus plants. Ordinarily, this would bring things under control pretty quickly. But imagine 4 square meters of plants. The gardener kills off 1 square meter, but that week the remaining 3 then double in area, to 6 square meters. The next week, with even more effort, the gardener kills off another fourth, 1.5 square meters, but the remainder doubles in area, to about 20 square meters. The hapless gardener is thus facing the very same type of exponential growth that occurs without his efforts; he's only slowed things down a bit.

Inflation is just the same. If, during one doubling of a space's volume, a process can end inflation in only less than half of the inflating space, then overall inflation can *never end*. This scenario is easily possible. One well-understood mechanism by which inflation can end is via the formation of a "bubble" (literally, a little ball) of noninflation. Such a bubble can arise spontaneously by a quantum process, after which the bubble grows, eating into the inflating volume outside. Despite this expansion, though, it can be shown that a given bubble eats up only a fixed fraction of the inflating space. Thus, like the lotus-covered region of the lake, the inflating volume increases exponentially—just at a slower rate than if the bubble-forming/inflation-ending process were not operating. And just like the lake, the inflating and noninflating regions, over time, build up a complex pattern[13] of space-time in different states (see the figure below). But unlike at Lake Dal, there is no limit to how long this can go on. Inflation never runs out of space to inflate into because it is creating its own space as it goes! This never-ending process has been dubbed *eternal inflation* or *everlasting inflation* by Andrei Linde, Alan Guth, Alex Vilenkin, and others who first understood it.

So, how does the observable universe fit into this view? Each region in which inflation has ended constitutes a huge, nearly uniform, fireball-filled volume of expanding space that can continue to evolve just as the big-bang theory predicts. Thus, everything we see cosmologically could be fit well inside such a region. Yet, other such

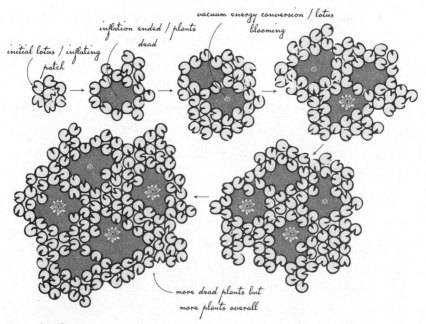

Eternal inflation.

regions exist, out there, separated from ours by domains of inflation. How many such regions are there? If inflation continues forever, at each time with more volume, and each piece of inflating volume eventually spawns a postinflationary fireball-type region, then it is quite clear that there is an endless number of these patches—each far larger than our observable universe!

What is less clear, but true, is that a sort of balance comes into being between the genesis of postinflationary regions (which remove inflating space) and the generation of new inflating space via the exponential expansion. These two competing effects lead to a situation that, when viewed from a wide enough perspective, looks more or less the same—regions of inflation punctured by regions of noninflation—at any time. With as many such times as you could ask for.

OUR CONCEPTION of the Universe has vacillated through history and across societies between a finite, fairly tidy affair with a definite start point, and an eternal sprawling, humbling profusion. The ancient Indians argued about it without conclusion. The Greeks undoubtedly did too. Kant proved both possibilities to be both necessarily true and impossible. In the twentieth century we had a face-off between the big bang and the steady state, with the steady state vanquished by a mountain of observational data.

But the same data, pointing to and supporting inflation, suggests a late-stage twist in the cosmic game: the Universe, it seems, may have chosen *both* paths.

Big bangs stitched together into a steady state.

32

THE PAINTING IN THE CAVE

(NEAR GANDEN, TIBET, 1613)

To call it a workshop, much less a studio, seems deceitfully generous. It is a cave, and you have developed a healthy fear of caves. But the tripa was adamant that you visit this supposed master at work.

You have to admit that the paintings display tremendous skill and patience in their craft—but their subject matter is baffling. Fierce demons, myriad Buddhas, tortured souls, turning wheels—it's all too much.

"So many realms!" says a soft but startling voice. "Each with no edge, no center, and everywhere circumference. What we think fixed and constant, changes, dreamlike, among them. Some are heavenly, some hellish. Some are so dark that light does not even exist. Some are so light that nothing else does. Some contain sentient beings; all are created by sentient beings. But one dharma pervades them all. And between the realms is a timeless sea of vast and seething energy. It is nothing, but it is everything, and it contains the seeds that birth all the many realms. It is beginningless and endless. It is boundless creation of time and space."

And as you look at the paintings, a wondrous, and terrifying, vision takes shape. But of what?

Thinking back on your conversations with the tripa, you realize it: "Ahh! This is a beautiful metaphor for how our mind operates. Throughout our life we move among the realms, which our mind creates. We make our own heavens and hells, our own very reality. But underneath this ever-changing mind is an unchanging pure awareness. So this is a picture of the whole interior universe!"

The painter looks at you impatiently, almost annoyed. "No metaphor," she says. "No interior. And not universe. Don't think so small."

Multiverse, a hypothetical collection of potentially diverse observable universes, each of which would comprise everything that is experimentally accessible by a connected community of observers. The observable known universe, which is accessible to telescopes, is about 90 billion light-years across. However, this universe would constitute just a small or even infinitesimal subset of the multiverse.

—Yours truly, in *Encyclopædia Britannica*

Like the painting in the cave, the idea of eternal inflation suggests that in both space and time, the Universe as a whole may be vastly, hugely, dizzyingly larger and more complex than the already immense universe that we can observe. And not just huger in size, but perhaps also in *diversity of properties*. Although there is a lot going on in the observable universe, part of what makes it comprehensible is that some things, like the basic laws of physics, the fundamental constants, and so on, are the same throughout. But if we take the (much) wider view afforded by eternal inflation, this does not mean they are the same *everywhere*. The painter's description sounds like poetic metaphor, but it is actually, for the most part, a matter-of-fact and fairly precise description of the sort of "multiverse" that eternal inflation might generate if variations in even these basic properties were included. Let's do some exegesis on it.

"So many realms! Each with no edge, no center,
and everywhere circumference."

These realms are the regions in which inflation has given over its
energy to some other set of fields—perhaps photons and other particles, as in our observable universe. Because each realm is preceded
by inflation, the realms are as smooth and uniform as Ali's globes.
They are thus well described by the *cosmological principle* underlying the big-bang model. To quote the painter and her contemporary
Giordano Bruno, their "circumference is everywhere, their center
nowhere." Each realm is plenty big to contain our entire observable
universe. (How big? We shall discuss that with Galileo soon.)

"What we think fixed and constant, changes,
dreamlike, among them."

The fine-structure constant α, describing the strength of the electromagnetic force, appears hard-coded into the universe. Observations
of intergalactic gas clouds billions of years ago and far away show
a value that is indistinguishable from here and now. In THROUGH
THE LOOKING GLASS we contemplated what might happen if constants like α were different in a speculative sense, but what would it
mean for a number like this to *actually* vary from place to place or
time to time?

There is an elegant set of ideas in which to make sense of it.
Within a decade after Einstein formulated general relativity, The-
odor Kaluza and later Oskar Klein developed a beautiful idea that
appeared to unite Einstein's gravity with Maxwell's electromagnetism. Kaluza and Klein supposed that our world actually has *four*
rather than three spatial dimensions, along with time. The fourth,
they assumed, is "rolled up" into a tiny circle so that at each point in
our three-dimensional space, there is, in fact, another *hidden* direc-

tion you could go, if you were tiny enough—though in going that direction, you would soon end up where you started.

This is a strange idea but has a surprising effect: Kaluza and Klein showed that Einstein's equations applied to this four-dimensional world are equivalent (if you gloss over the tiny scale of the curled-up dimension) to Einstein's equations *and* Maxwell's equations operating in the three other dimensions. In this correspondence, very interestingly, the value of the fine-structure constant α in Maxwell's equations is directly related to the *radius* of the tiny circle. Moreover, nothing prevents this radius from changing from place to place.

In this theory, α thus goes from being a constant of nature to being a *field* that can vary, and even have dynamics like the electromagnetic or the inflaton field. This demonstrates that even "fundamental" constants can, in principle, vary across space-time.

While the Kaluza-Klein theory has fatal flaws, it was reincarnated in the 1980s and 1990s in string theory. In string theory there are also (by assumption) three *large* spatial dimensions, but also necessarily six or seven *small* dimensions, wrapped up like Kaluza and Klein's circles, on a tiny scale. But rather than being just a simple circle, the extra space could have a very, very complex geometric structure. And rather than requiring just a single radius, it could take hundreds of parameters to describe the structure of the hidden space that resides at each point in our three-dimensional world. These hundreds of parameters then represent *hundreds of fields*— that is, numbers that can smoothly vary from place to place—that determine not just α, but many of the other constants that appear in the standard model of particle physics.

If this picture is right, it provides a mechanism by which what seem like the basic properties of our universe (lowercase *u*) could be quite different in other regions of the Universe (uppercase *U*!), or what is often called a *multiverse*. Further, inflation, if it is operating, could take a tiny region in which the extra dimensions are arranged in the same way, and inflate it up to far larger than everything we can observe, so that the properties *appear* uniform everywhere. Yet elsewhere, beyond the reaches of our observations, other universes that are just as large could have wholly different properties governing the macroscopic world.

> "Some are heavenly, some hellish. Some are so dark
> that light does not even exist. Some are so light that
> nothing else does."

What sort of properties could vary across the universes? Even if string theory were completely misguided, cosmological properties such as the amount of dark matter, the density of the universe at a given temperature, the nonuniformity amplitude, and so on might, in principle, vary, as all might have properties governed by fields that could vary from domain to domain. It is not hard to imagine a universe that is essentially all radiation—light—with nothing like matter ever forming in appreciable quantities, or doing anything interesting.

If something like string theory *is* true, then so-called fundamental constants like α could vary. Moreover, it may go beyond the constants. Electromagnetism, for example, is part of the standard model of particle physics and is distinct from the other forces (weak and strong) at low enough energy scales. Like the fundamental constants, the properties and even existence of various particles in string theory are governed by the compact dimensions. We can then imagine a different geometry for these tiny dimensions that corresponds to a modified particle physics model *without* electromagnetism. Just as someone blind from birth does not "see black," but rather has no sense of seeing at all, such a universe would not just be dark: it would *have no light* at all.*

---

* Electromagnetism serves all manner of extremely useful purposes, so in creating a universe like ours, letting there be light is indeed a nontrivial, and fairly well advised, first step.

"Some contain sentient beings;
all are created by sentient beings."

Life like us, based on chemistry between molecules held together by electromagnetic forces, could obviously not exist without electromagnetism; that would be a disappointing universe. Likewise, we've seen that even if we do not inhabit the "best of all possible" worlds, ours is quite nice: it has many habitability-providing properties that could easily have been different, if we are imagining basic physics varying from one universe to another. So it seems very reasonable to suppose, per one of our solutions to the problem of theodicy, that only *some*, and perhaps very, very few, of the universes brought into being by eternal inflation would contain sentient inhabitants at all. The rest would seem to have nobody in them wondering: "What's it all about?"—or even wondering, perceiving, thinking, observing, or acting on anything whatsoever. One might even ask whether such a universe can meaningfully be said to exist at all.

"But one dharma pervades them all."

A crucial aspect of the string theory picture, with its varying constants and laws of physics, is that *not everything varies*. Rather, this view posits that the fundamentals of quantum mechanics, general relativity,[14] and their application to strings and similar objects, are truly universal, functioning in essentially the same way across all space and time. What varies is the particle and field content, including the fields that determine the so-called fundamental constants, like α. Different "laws of physics" apply in large regions in which the ambient energy has dropped low enough that these fields are generally in their resting, or *vacuum*, state. Between regions with different vacuum states, however, the fields must traverse high-energy areas in which the fields are *not* in any vacuum.

In this picture, then, one overarching set of physical laws—one dharma—holds at very high energy scales, and coexists with many, many, many "choices" of detailed physical laws and constants that pertain to lower energy scales.

> "It is nothing, but it is everything, and it contains
> the seeds that birth all the many realms. It is
> beginningless and endless. It is boundless creation
> of time and space."

In this way a single set of physics—quantum mechanics, general relativity, and inflation—can give rise to a multiverse that extends indefinitely forward (and possibly backward) in time, unboundedly in space, and is potentially (if something like string theory is correct) populated with a dizzying diversity of properties. If this is so, then

> "Not universe. Don't think so small."

speaks for itself.

33

A DIALOGUE CONCERNING INFINITELY MANY THINGS

(PADUA, ITALY, 1608)

"I assume," begins Galileo, "that you know which of the numbers are squares and which are not."

"I am quite aware," you reply, "that a squared number is one that results from the multiplication of another number by itself; thus 4, 9, and so on are squared numbers that come from multiplying 2, 3, and so on by themselves."

"And would you say," he continues, "that there are more numbers than squared numbers?"

"Clearly," you answer. "Since there are numbers that are not squared numbers, there must be more."

"And yet," Galileo muses, "if I begin to count squared numbers, I can use all numbers to do the counting. Watch: 1, 2, 3, 4, . . . , turns into 1, 4, 9, 16, . . . ; thus, for each squared number I have a regular number that is its root, and for each number I have a squared number. When there is a perfect one-to-one correspondence between sets of objects, I would ordinarily say that each set has the same number of objects, would I not?"

"I'm quite perplexed," you reply, "for while I agree with your argument, I feel that I could also ask what fraction of numbers between 1 and 10 are squares. It is 3/10. Yet between 1 and 100, only 1/10 are. And so on: as I increase the size of my comparison set, the fraction of squares goes to zero. So it seems that the relative number of squares and numbers depends on how you count! How confusing!"

Galileo nods. "So, what do you conclude about the relative numbers of numbers and squared numbers?"

You ponder a bit before replying: "It seems we can infer only that the totality of all numbers is infinite, that the number of squares is infinite, and that the number of their roots is infinite; the attributes 'equal,' 'greater,' and 'less,' are not applicable to infinite, but only to finite quantities."

"It's bewildering—maddening even," Galileo replies, almost despondently. Then he suddenly recalls something. "Remember that new device for seeing small things? It arrived today. Let's take a look. Perhaps I will leave this matter of infinity to drive someone else mad instead."

> These are some of the marvels which our imagination cannot grasp and which should warn us against the serious error of those who attempt to discuss the infinite by assigning to it the same properties which we employ for the finite, the natures of the two having nothing in common.
>
> —Galileo, *Dialogues concerning Two New Sciences*[15]

The concept of infinity has both inspired transcendent thoughts of the divine, and driven mathematicians to madness. At least as far back as Aristotle, people have grappled with the idea of numbers that have no end, and with how to think about a corresponding *infinite set* of things counted by those numbers. Much of this early thinking related to the debate—continued to the present day—regarding potential versus actual infinity. One might consider *potential* infinity to be what is latent in the statement "keep counting": there is no limit to how high you count, but you can never actually attain infinity. An *actual* infinity would be an infinity that is somehow realized all at once. A reasonable position to take might be that infinite sets exist as mathematical objects (though some mathematicians deny this), while the actual physical world can contain only numbers that are arbitrarily high, but not actually infinite. As we will see, however, some physicists in turn deny this.

The mappings that, in his *Dialogues*, Galileo constructed between integers and squared integers are an excellent place to start. As one of the earliest cogent explications of the paradoxes of infinity, they demonstrated two things.

First, counting becomes ambiguous. If you are comparing two *finite* sets of objects to see which has more objects in it, the order in which you count the objects does not really matter. But for infinite sets, by deftly choosing your order you can obtain many different answers—for example, that there are as many squared integers as integers, or six times as many even numbers as odd numbers. Thus, many sets that might *seem* to contain different numbers of elements can be said to really be "just as big" as one another[16] in a way that is quite different from sets of any finite size, no matter how large.

Second, the dialogue indicates that we can measure the relative frequencies of different types of elements in a set, *as long as* we have the set arranged in some prescribed manner. That is, if we order all of the integers in, well, order, then we can count relative numbers of even and odd integers up to some cutoff integer N. Then if we increase N unboundedly, we see that the ratio between even and odd integers approaches one to one. Likewise, the ratio of squared integers to integers approaches zero. Now, of course, had we ordered the integers differently, we could have obtained a different result in either case, but there does seem to be a somewhat "natural" ordering to the integers, suggesting something a bit more special about the ratios we obtain in using it.

NOW LET'S RETURN to cosmology. If eternal inflation is true, then the Universe—or rather multiverse—goes on forever, spawning an infinite number of postinflationary regions. And if that's true, we have something to worry about. If we ask: "As inhabitants of this multiverse, what properties are we most likely to observe?" then what we are really asking is something like: "Given regions with many different properties, what type of region is most common?" or, alternatively, "Given that I am in some *arbitrarily chosen* region, what is it most likely to look like?" These are questions about relative frequencies of properties in an infinite set, just like "evenness" and "oddness" and "squareness" in the integers. And just the same issues can arise. Inflation, though, goes a bit farther by adding an additional ambiguity on top of (though in some cases interchangeable with) the counting ambiguity, which is the *time ambiguity* of general relativity. As we saw when RELEASING THE DJINN, a given space-time can be sliced up into surfaces of simultaneity in many different ways. Each is just as good as the others in principle, even if in practice some are much nicer than others. This ambiguity in what counts as a given time, combined with the infinite extent of an eternally inflating space-time, leads to some pretty baffling problems.

Suppose there are two types of regions in the Universe: enlightened regions and unenlightened regions. Regions are unenlightened by default, but here and there in the cosmos a teacher appears, and the regions that this teacher's teaching reaches become enlightened. Let's also assume, for simplicity, that the teaching is spread at the speed of light (say, via enlightening broadcast). Then the eternally inflating Universe will look overall like a hodgepodge of enlightened and unenlightened regions, not at all dissimilar from Lake Dal's aspect of lotus-filled and lotus-free regions.

Now, we might ask: "How much of the Universe is enlightened right now?" But it is not hard to show in this circumstance that depending on how you define *now*, the answer might be that almost all of the Universe is enlightened, or almost none of it is! Your definition of time can be very much like the way you "order" the integers, and just as you can order the integers and then count them so as to find many more odds than evens (or vice versa), you can define "now" to do the same with enlightened or unenlightened regions.

It might seem that it would help to ask how much of the Universe *up until now* consists of enlightened regions, but vexingly, it does not help much. A hallmark of exponential growth in time is that the most recent epoch contains just about as much as all earlier epochs—so the overall ratio of enlightened versus unenlightened volume is dominated by "now," however it is defined.

These issues and ambiguities may leave us wanting to scurry for the cover of the finite. Can this be done? For example, suppose we limit ourselves to just *our* universe. Is that finite? Well, the *observiball* universe back to some big-bang cosmic time is indeed finite. But what about the full spatial region at that same time, which shares the postinflationary properties of our universe? Is that finite?

Nope! Consider a single enlightened region of space-time (see the figure below). Since we've assumed that enlightenment grows at the speed of light, this region fills the future light cone of the point at which the teacher started teaching. Now our freedom to flexibly define "now" lets us do something amazing. Examine the curve just inside the enlightened region in the figure. Because it always goes more horizontal than vertical, it can be considered "space" at a single

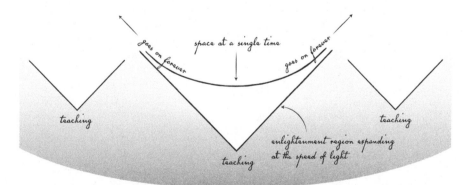

Enlightened (light) and unenlightened (dark) regions of an inflationary space-time, if the enlightening words of a teacher spread at the speed of light.

time. And because the enlightened light cone goes on forever up the page, so does this space. So this is an *infinite* space. And it's nestled inside what looks like a *finite* growing region. In this sense, relativity allows a single teacher to enlighten an infinite volume, all at once![17]

The structure of our universe, if it is within a bubble of noninflation, turns out to be almost identical. From the outside, our bubble would look finite and expanding, but from the inside: spatially infinite! These features turn out to be quite generic for inflation-ending surfaces in eternal inflation.[18] It is therefore a quite universal property of eternal inflation that it spawns infinitely many postinflationary universes, and also that each of these can be described as spatially infinite inside.

There is no easy escape from the infinite. It can be maddening.

If we inhabit a multiverse like that described by eternal inflation, what explains the properties *we* see around us? Are they common? Uncommon? Common among the best of all possible worlds? The best of all common worlds? Who are "we" anyway?

Nobody knows.

34

SICKNESS UNTO DEATH

(ON A ROAD IN CHINA, 1615)

The fever hit first, then a searing headache. You could not open your eyes, let alone stand.

So you lie in the small room as darkness encroaches. The room is in . . . where? Tibet. No, that was months ago. Where?

You cast your mind back, and . . . nothing. Nothing! Too tired, the road is a fog. No: there was a gondola, and a ship, and desert, and a lake, and, and . . .

Your mind is a sea of blinding light with waves of pain. Your body fades across the horizon.

You are . . . where? You remember . . . a desert.

You are . . . who? A student, a traveler, a thinker. A traveler. A thinker. A thinker.

You are someone. Who, though? You think; and then you cannot.

Blinding darkness, your breath. You are here, there. Here and there.

You are, and then. Are.

Are.

Are.

A single note plays, and plays, and fades, and fades, and

. . .

Feel.

You feel!

You are, you think. Who is thinking? You are, you remember. Body, you have one. So tired. You remember, a student, a traveler, a desert, a ship.

Bumps, sounds. You are in motion.

Your eyes open. The inside of a cart, on a road. A voice: "Shenyang."

So tired, you sleep.

Suppose that this great Earth were totally covered with water and a man were to toss a yoke with a single hole into the water. . . . And suppose a blind sea turtle were there. It would come to the surface only once every 100 years. Now what do you suppose the chances would be that a blind turtle, coming once to the surface every 100 years, would stick his neck into the yoke with a single hole? . . . just so, it is very, very rare that one attains the human state.

—Shakyamuni Buddha[19]

You are going to die, at some unknown time in the future. We've already covered that. But while the date of your death is uncertain, it can be predicted *statistically*: each possible number of remaining years can be assigned a probability. Insurance companies have quite accurate methods for this calculation, known as *actuarial tables*. Very roughly, most people live about 75–80 years, so you could just subtract your age from this number to estimate how much longer you have. But you can do much better. If you're a regularly exercising, nonsmoking woman, your number will be considerably higher than for a sedentary, smoking man, and by taking into account gender, current age, health, various risk factors, lifestyle, and all manner of other data, you could get a much more tailored prediction.

We can look at these tables as a sort of *predictive theory* about life spans. When you consult the theory, you ask: "Given that I am a typical *X*-year-old, how much longer do I have to live?" You could, if you like, imagine all of the *X*-year-olds in the world, including you. Given only information about age, the theory spits out a result that applies to all of you. But if you like, and if the theory can make use of it, you might also specify your gender to reduce your *reference class* of similar people by about 50%. Likewise, each additional piece of information you provide trims your reference class down, and makes the predictions more specific. Of course, you can be *too* specific; theories are limited, and no plausible model can usefully fold in, say, the amount of mayonnaise you use or how careful you are in parking lots. And if you specify a whole lot of information, such as zip code and age in days, you'll narrow your reference class down to a single person without giving the model anything more useful to work with. Nonetheless, in general, given a very good theory, the more information you specify, the better predictions you will get. And this echoes what we do a lot in science: assume a theory, gather as much pertinent data as possible about the world, feed it into the theory, and obtain predictions.

However, folding in more information may not *always* be better, particularly if you are trying to *evaluate the theory itself* in terms of how good its predictions are. Consider whether you are, right now, dreaming. Are you?* That's a theory you might want to evaluate in comparison to an alternative theory that you're awake. In doing so, you're *not* going to want to simply accept any previous observations you've made and use them to make future predictions. You'll want to *check* those observations themselves against the theory, in particular for internal consistency. Do you remember how you got to where you are right now, or are there any gaps in a coherent timeline? Did

---

* In my experience, most times you ask yourself whether you're dreaming, or say to yourself: "Whoa, that's strange!" you are dreaming. So it's a pretty good bet in the present circumstances.

anything strange recently happen? Do your present surroundings surprise or perplex you in any way? All these questions can give useful clues, and if any of them indicates a potential dream state, you could do some tests, such as testing the stability of text (easy, since you're reading this), jumping to test your weightiness (careful of the cliff!), and so on.

Accepting certain data in order to make predictions is often called *conditioning on* that data. But when you condition in this way, you lose the ability to discriminate between theories that might have predicted different things for that data. In terms of reference classes, putting yourself in a very specific class is useful for getting very well tailored predictions. But if you blithely accept that you are part of a reference class that is extremely unlikely (say, people being stalked by monsters), given your theory (that you are awake), you may fail to realize important truths (that your theory is wrong, and you are dreaming). Rather, you want to assume *less*: to assume that you are some sentient being that can potentially dream, as well as reason, observe, and evaluate, and compare your observations—both past and future—to your two candidate theories.

The question of whether you are dreaming (which you've hopefully resolved by now) may not seem like the most central or germane. But is it really so different from the existential questions that arise as we contemplate ourselves as beings, among who knows how many different types, awakening in a Universe of myriad possible properties, in a reality of a fundamentally unknown nature?

The idea that we should reason as if we are a *randomly chosen member* of some reference class is often known as the *self-sampling assumption*.[20] In a sense, this idea encompasses the way we often do science, since the reference class might be defined as "experimenters who have set up an experiment exactly as I just did." It can also, though, apply to murkier questions, like "Why does the observable universe appear fine-tuned for life?" If the reference class is, say, "humanlike observers," then as long as the Universe (or multiverse) as a whole admits such observers at all, random members of that class will see a universe that looks fine-tuned for life; universes that do not look fined-tuned for life have no such observers in them! This is just another way of framing the multiverse explanation of theodicy.

Once a reference class is assumed, considering ourselves *randomly chosen* members of that class seems somewhat obvious: what *else* would we do other than assume we are a random sample from it? Indeed, to do otherwise would point to something important and relevant that differentiates us from other members of the reference class; in this case we really should just define a smaller reference class that takes these into account, and assume we're a random member of that. In this sense, the big issue is this: Exactly which reference class do we assume when we reason about what we see when we look out at the universe?

That is, let's imagine that we've opened our eyes somewhere in a multiverse of infinite size and tremendous diversity. What should we expect to see? And is it what we actually see?

Well, what is the reference class that we define when we say "we"?

Is it simply a sentient being with some level of awareness?

Is it a self-aware, thinking being?

Is it a humanoid, carbon-based, intelligent life-form?

Is it a life-form advanced enough that it ponders questions about the universe such as the fine-tuning of physical constants?

Is it an actual human, living in an advanced society, who likes to read books about perplexing cosmological conundrums?

Is it someone just like you, with all of your memories and perceptions?

For a given model of a multiverse, these reference classes provide a sort of spectrum of how much to condition our predictions on, and each would provide a potentially different answer to the question "What do we expect to see when we awaken?"

A mere sentient being might awaken to find itself an ant or the like; that's probably too little conditioning. A self-aware thinking being? How many types of those might there be, and what might they be thinking? A human in an advanced society might wake up on a road to Shenyang. Someone just like me woke up just like me— but what about you?

What is the right thing to do?

And if what we expect to see depends so much on who "we" are, then does the universe make you, or do you make the universe?

35

AN HONORED GUEST

(AGRA, INDIA, 1611)

The wedding of Jahangir and Nur Jahan, with elephants, fire-works, palanquins, and an astonishing amount of gold, is more opulent than anything you've ever seen—or even consider possible in your native lands. Or at least, this is how it appears from a great distance, which is as close as you can get.

After two years of hard traveling, you suddenly and powerfully miss the comforts of home, even pining, for a moment, after even a tiny fraction of the wealth of the Mughal royalty. You are feeling poor, and insignificant.

You notice a rag-dressed but bright-eyed sadhu who has been peering at you as you've watched, and seems to read your mind: "So much gold, so many things, so much comfort. You want it, don't you?"

"Sometimes," you accede. "Sometimes I think it would be nice."

The sadhu nods: "But all of this!" He points, and you follow his gestures to the elephants, the horses, the people, the embroidery, the dust-covered ground, the running fountains, the flowers; then the music, the sound, the breeze, the trees; then the lands, the mountains, the stars; then history, society, humanity; then the ebb of time, to beauty and tragedy, to thought, belief, truth . . .

"All of this," the sadhu insists, staring into you, "it is a wealth all yours, made by and for you, existing only with you."

And you see it. He is right.

In the perspective of the violences of matter and field, of these ranges of heat and pressure, of these reaches of space and time, is not man an unimportant bit of dust on an unimportant planet in an unimportant galaxy in an unimportant region somewhere in the vastness of space? No!

—John Wheeler, *Preface to* The Anthropic
Cosmological Principle

Starting in Galileo's time, humanity's sense of its special place in the Universe has taken a series of blows. Earth was at the center of everything, with some rather insignificant stars and planets circling, and the history of Earth was more or less that of recorded human history. But that is not so. Earth is billions of years old, to humanity's millions and history's thousands. The Sun we orbit is one of hundreds of billions in our galaxy, which is one of hundreds of billions in the visually observable universe.

The vastness of this observable universe is almost beyond comprehension when compared to the Earth. To Earth's mote of dust the Sun is a grain of sand, and our galaxy is a metric ton of sand arranged with the stellar grains 10 kilometers apart. But if this tiny model of the Milky Way is scaled down, down, down, down to the size of a snowflake, among the galaxies of our observable universe the Milky Way is just one flake of all those falling over a Himalayan valley on a snowy evening.

Moreover, we have good reasons to suspect that our observable universe is just one of enormously or infinitely many such regions stretching unboundedly in space, time, and possibly other dimensions.

In this, the West has merely caught up (albeit with greatly more precision and concreteness) to the classical worldviews of the East, with cosmic cycles of vast age spawning innumerable lands and beings across myriad planes of existence.

It's a conception that is truly awe-inspiring, and profoundly humbling.

And yet . . . we are quite special. We have poked our turtle nose into the right floating yoke to be an honored guest at the cosmic festival. We are composed of a quite special type of material—elements forged in massive stars and their explosions—that makes up only about one part in 100,000 of the universe's density. But crucially, this matter supports complex chemistry: there are a staggering number of ways atoms can form molecules and those molecules can interact, enabling an amazingly rich set of phenomena wholly lacking in other cosmic constituents, like dark matter, dark energy, ionized gas, neutrinos, and so on. And we're made of enough atoms for this complexity to fully play out. Each of us contains about 10^{28} particles, allowing molecules (like DNA) with billions of atoms in them, cells with billions of molecules in them, and creatures with billions of cells. We're also small enough that chemical signaling can support many billions of coherent "mental" actions over a year of life, and a billion such lives over an evolutionary timescale.[21]

And yet . . . all of these considerations are just the physical scaffolding to support what *really* makes us special: the ability to perceive, to learn, to evolve, to predict, to experience, to feel, to think and wonder. Unless it is doing it very stealthily or in some very different way than we imagine, nearly none of the matter in the universe is doing this.

But it's arguably the most important thing that the universe does. Many millions of cold rocks are floating around in the Oort cloud surrounding the solar system. Do any of them matter? Would the universe be better, or worse, or in any meaningful way different if those rocks did slightly different things? If the galaxies were arranged differently in the cosmic web, or all the neutron stars in the universe were permuted in their positions, would it matter? In contrast, we very much care about the things that thinking, feeling beings do. Their welfare matters. The suffering and joy of a single person matters. And really, if you think about pretty much any goal, or aim, or ranking of states of affairs, you will find that at the bottom it is connected to the *preferences of some thinking, feeling, conscious beings.*

Put another way, suppose you were given the unbelievable power and ability to create an entire universe—say, 10^{80} special particles that would fit on a grain of sand. You might, THROUGH THE LOOK-ING GLASS, see the rather featureless start of this universe before it evolved and grew in complexity. But suppose it were guaranteed that neither you nor any other conscious being could ever look at the universe after its first moment, and that *none would exist inside it.*

Would there be any point to making it?

THUS there is an argument that only universes like the one we inhabit, with aware beings inside them, *matter.* Is there also an argument that only universes like the one we inhabit *exist*?

That sounds crazy. But let's take a look.

The investigation of THEODICY suggested curious connections between ourselves as complex thinking sentient beings, and the properties of our universe. All of the different ways of thinking about that connection suggest something quite amazing going on.

After being duped by coincidences in the port town of Tripoli, we explored one fairly compelling explanation for fine-tuning: that our "Universe" is really a vast and diverse multiverse consisting of inhabitable regions quite like ours, potentially some habitable regions quite unlike ours, and many, many boring and uninhabitable regions. The possibility of such a multiverse was grounded in eternal inflation, with its ability to endlessly spawn new universes.

And there is another way. Suppose that inflation is *not* eternal. The Universe somehow begins, inflates for a while, then stops. Seems simple, but this is a *classical* description; let's also include quantum mechanics. Then the initial state of the Universe, whatever it is, evolves not into a single later Universe, but a *superposition* of many possibilities. These superposed universes could have quite different classical properties, in a manner not dissimilar to eternal inflation–spawned universes. Because universes are big (if anything is!), their histories would naturally decohere into alternative, classically describable worlds. If we take seriously the ontic view of quantum theory, in which each of these worlds is just as real as the others, then we have what can be called a *quantum multiverse*: not just many worlds, but many universes. Much as in the inflationary multiverse, physics may allow diversity of properties in the fundamental constants, cosmological properties, and so on, and in a similar way support the explanation of theodicy.

In either version of the multiverse, this adds up to a pretty radical view: the Universe matters because it is *so big and complicated* that it almost necessarily has an infinitesimal part within it that can support things—beings and minds—that make it matter.

We might be even more radical, though, and ask: "Why countenance all the waste?" Should we actually count as "real" all those other regions and branches of the quantum state, if nobody is home?

This question leads to two simple but dizzying others:

What does it mean for *us* to exist?
What does it mean for *something other than us* to exist?

PART 5

Who Am I?
Don't Know!

To study the Buddha Way is to study the self.
To study the self is to forget the self.
To forget the self is to be actualized by myriad
  things.

—Eihei Dōgen, *Moon in a Dewdrop*

To be oneself, simply oneself, is so amazing and utterly
unique an experience that it's hard to convince oneself
so singular a thing happens to everybody.

—Simone de Beauvoir

36

WHO SLEEPS, PERCHANCE TO DREAM?

(UNKNOWN)

You are at home, in Padua, at university, at an exam for which you have not studied. Of course. Dreams are often this way.

The perspective shifts, and you sit at a low table, facing a scroll of indecipherable text, upon which you cannot seem to focus.

You are walking over a bridge in Venice, watching the gondolas pass underneath. What scroll? What exam?

A glowing light leads your attention as your world shifts to an age of marvels, where words and moving paintings appear on phantasmagorical, ever-shifting tablets. The text reads: "Sta volando verso il tuo cuore."

You dream of moving through a labyrinthine jungle, chased by an implacable, silent, stalking beast.

You awaken, relieved. But the relief returns to panic as you find you cannot move, but for the swaying of the small cart you ride in.

You awaken to the swaying of the ship, and it all returns to you: the cart, the beast, the bridge, the exam.

You awaken. You were in a ship but now lie on a hard pallet in a cold room. A phrase about an arrow burns in your mind, almost forgotten.

Who was it that dreamt?

If the self were the aggregates
It would have arising and ceasing as properties.
If it were different from the aggregates,
It would not have the characteristics of the aggregates.

—Nāgārjuna, *The Fundamental Wisdom of the
Middle Way*

Who is it that reads these words? Out of all the beings through space and time, there's one very particular one, *you*, who is experiencing this. *You* had some particular parents, the childhood you remember, the day you had yesterday; and *you* drifted off to sleep last night.

And when you awoke this morning, if you are like most people you very quickly jumped into your same life, rising from your bed, thinking about the day to come, remembering what you did—or left unfinished—the day before.

It wasn't a seventeenth-century explorer, or a horseman in Manchuria, or a pioneer on Mars, who went to bed and awoke as you; no, it was *you* who went to bed. You, the actor, the experiencer, the singular protagonist of the life you experience from the *inside*, the one who thinks your thoughts, makes your decisions, and holds their responsibility. *You* went to sleep last night.

But how do you know?

You might object that this is obvious: *you* are the one who remembers yesterday, and the weeks and months before. A whole lifetime is stretched out and instantly at hand for your inspection, even if some of the memories are more clear than others. You feel comfortable in your body, used to how everything works. The world around you is familiar, understandable, somewhat predictable. You are *you*. You feel it, with the utmost conviction. It cannot be doubted!

Can it?

Imagine that "You" are a little nugget of soul stuff, unitary and inviolable, abiding in your body and pulling its strings. It often feels that way, does it not? You think of "your body" and even "your mind." It feels that there is something separate, one step back and higher up, that is the owner, some little essential self of You. This You is singular, at the center of it all.

But what, then, is the *relationship* between this You and the things you identify with: your body, your sensations, dispositions, thoughts, perceptions, and so on? Is the You just composed of these and other things, existing insofar as they do, or as some particular combination of them? If so, it must begin and end with them, evolve with them. As Nāgārjuna says: "If the self were the aggregates,* it would have arising and ceasing." But the You does not feel like a partial, evolving, constructed thing; it feels *binary*: you are You—not halfway, not sometimes, but only and always You and nobody else. It does not at all feel like something to be lumped in with a bunch of always-shifting and ever-evolving properties and processes. The You feels distinct, unitary. There's a singular subjective view that you feel your You has been attached to for your entire existence.

And yet, as Nāgārjuna goes on: "If it were different from the aggregates, it would not have the characteristics of the aggregates." That is, if the You really were singular, distinct, separable in some way from all the processes in your mind, some qualitatively different sort of "substance," then it could be *detached* from all those things. Could You not then flit to some other body? Could You not have flitted from some other body as you slept? If You had, what would you then remember? It is fairly clear that memories are kept by the brain, as are the connection with your body, your skills, your ten-

---

* These aggregates were often listed as physical body, sensation, perception, dispositions, consciousness, and cognition, but there are other similar lists.

dencies. Had You switched identities in the night, you would never remember that previous life, but simply and immediately assume— no, *know without a doubt*—that You had always been living this one. Spend a few moments supposing that you last went to sleep on a sailing ship, after awakening to a surprising stillness. Really try.

How do you know You didn't? And if you can't tell, of what use is that sort of You anyway?

When we think of everything we identify with, it's easy to see that much of it was accumulated through our time here on Earth: memories, capabilities, patterns, habits, and freedoms. It's just as easy to see how these could be lost: as memories fade, years of disuse dull abilities, new actions break routines. We have, perhaps, seen them lost to those ravaged by disease or suffering from traumatic injuries. We can imagine them removed one by one. But if all of them are taken away, what is left? Still that nugget of You, you can't help but feel. It's still there! It must be!

But what, really, is it?

We clearly have a mental process that can *create* a feeling of personal identity, of coherent "self." Have you ever dreamt you were someone else, with a whole different history and identity? In a dream you are often cut off from most memory of your daily life. It does not strike you as incongruous to be in a situation wholly at odds with your waking existence. Crucial people—children, spouses, friends—may be missing, or extra, without causing any concern whatsoever. Given whatever happens, whatever you can recall, whoever is there, you effortlessly construct a story line that makes sense, and you generally experience it as happening to You, right? And yet, after several such dreams you might awaken to realize that none of those subjects were the one that wakes up, which somehow encompasses them all.*

---

* Makes you wonder who is going to wake up from this dream, does it not?

So if you can, in a moment, construct a sense of "You"-ness out of almost arbitrary raw materials, it makes that whole you-ness sense a bit suspicious, doesn't it?

Moreover, from a biological and evolutionary standpoint, it makes good sense that organisms would develop a sense of "self" to help self-identify as individual organisms, to consolidate information and understanding, and to provide a nexus for action that can imagine, plan, and decide so as to keep the body and genes alive in a hostile world. And modern neuroscience and physics have no call for, or even really space for, some sort of additional quasi-physical "substance" of which the "You" could be composed, interacting with but separate from the body. So, the doctrine of "no self," declared by the Buddha and passed down through Nāgārjuna and others, is quite consonant with a modern scientific view that the "self," the "You," is a somewhat ephemeral construct.

It's easy to say that. But it's the hardest thing in the world to really *believe* that the You, the essential self—the thing that *thought* itself "Descartes," and so *was*—is no more (or less) real than the thoughts, feelings, concerns, and perhaps perplexity going on right now in the mind called "yours."

Some people say they believe this but, like almost everyone else, value this "self"-ish thing above nearly all other things.

Some people, instead, are convinced of the inviolable nugget of selfness. But can they point to it, or explain the nature of the glue that would adhere it to awake day after day in the same body?

Who is it who is believing, or doubting, the You?

Who was it, just a moment ago?

37

A SIMPLE ARRANGEMENT OF SOME BITS

(DISTURBING ARABIAN CAVE, 1610)

You have had quite enough of the djinn and its antics, and you turn to leave the cave. "But," teases the djinn, "we are just getting to the best part." And suddenly, your surroundings shift, and looking about in astonishment, you realize that you have been instantaneously transported to an alcove somewhat deeper inside the cave. "What *happened*?" you implore the djinn.

"Ah," the djinn replies, "I simply took note of the relative positions and velocities of the atoms in your body, disassembled you, moved the atoms to the other side of the cave, and restored them to their original configuration."

"But I am not just an arrangement of atoms!" you object, but with sudden doubt, "am I? What of my inner life? My consciousness? My soul?"

"I only moved atoms," replies the djinn with a glint in its eye, "so you tell me."

You are quiet for a moment. "But," you then object, "you have proved merely that my bodily arrangement of atoms is necessary, but not that it is sufficient. I may inhabit my body, and move when it does, but that does not mean I—the unique, conscious, subjective I—am the same as it!"

Just as suddenly as before, you find yourself in a new location. "Stop that!" you exclaim. But as you say it, you hear a curious echo from across the cave, and with horrified amazement you see what appears to be an exact duplicate of yourself across the cavern, replete with an expression of horrified amazement.

"But that is not me!" you both cry in unison. Then, after exchanging a perturbed glance with your twin, and trying to ignore your doppelgänger as it says something similar, you go on: "You can create a physical copy. But my experience was perfectly continuous. How can that experience split?"

The djinn replies: "I see no contradiction: there was one, now there are two. However, I can see that I have caused you distress, which I will rectify by another simple rearrangement." And in a moment, you see your twin disappear, leaving only a mound of muck on the cave floor.

You are horrified, then relieved, then angry: "How could you create a living being—just like me—then go and kill it?!"

The djinn sighs. "Very well. I will be less cruel next time." You suddenly appear across the room once again (restored to your original position near the exit). You look with trepidation across the cave. But you see just a fresh, steaming pile of muck. "This time," explains the djinn, "I duplicated you, but before one duplicate was aware for even an instant, I decomposed it painlessly. Surely this does not trouble you."

"How pointless," you reply. "Now we're back to just what we did the first time."

But then you become uneasy. Are you back where you began, or not?

We can start by discussing the much-discussed case
of the man who, amoeba-like, divides.

—Derek Parfit, *Personal Identity*

What if some ultra-advanced technology or fantastical being could cause your body and brain to be taken apart, bit by bit, and reconstructed in the *exact same arrangement* somewhere else? Should you fear this process happening to you?

The question hinges directly on the nature of the "You" and whether there is something beyond the incredibly complex arrangement of particles and fields that make up your body (your body's physical "state") that would be missing, were that body destroyed and perfectly reconstituted. Even if it is hard to accept, let's suppose that there is *not*: let's suppose that whatever You are, you depend wholly on, and are in one-to-one correspondence with, your particular physical instantiation, with different mental states corresponding, even if in mind-boggling complexity, with different physical states.

Where does this assumption lead?

First, if this "teleportation" is possible—and harmless—then why not simply run the reconstruction process twice to get duplication? To be clear, there are reasons we have seen and will see, to think that such duplication may be either hopelessly impractical or even impossible. Such thinking would conveniently avoid some of the implications. But let us not be convenienced.

So, then, try to imagine it: If you are standing in the cave and the djinn blinks, *on which side of the cave are you next?* For the djinn, there is nothing confusing, any more than if you cast two copies of a bronze sculpture. But for you, there is nothing *but* confusion. Your very clear sense of "selfness," which continually affirms that you are a singular being that persists through time, is in stark contradiction to what the djinn sees. It's clear that you can't be on *both* sides in a meaningful sense: there is no way for "You" to experience both sides any more than you experience the sensory input of any other person; there is only one set of sense organs connected to each brain! But neither does the duplication process favor either side over the other: there is perfect symmetry between the copies, and both will declare with utmost conviction to the djinn that they are "You"; both will share every memory up until the duplication event, as well as every propensity, mental ability, ongoing thought, and so on—as long as these are, again, in one-to-one correspondence with the physical body and brain. So, there is no sense in which one of them can be You and not the other.

Our experience of the world is entirely subjective by definition, but we also believe that there are other such viewpoints held by others, who all agree on an "objective" world to boot. Yet we are wholly inexperienced with the idea of a subjective point of view bifurcating into two; it feels self-contradictory, like the center of a sphere turning into two centers of a sphere. A sphere only has one center!

It is difficult to really engage in this introspection without the rather disquieting feeling that the "You" is indeed not at all what we ordinarily feel it to be. (It is happening to me right now as I write this—a feeling of being unmoored from time and history, of suddenly arising at this very moment, poking my nose through the yoke.)

AFTER PONDERING that a bit, let's now ask our original question again, but of duplication: Should you be *afraid* to be duplicated? It would seem that if you can teleport unafraid, there is nothing obviously scarier about duplication, and perhaps there is something to be gained, if you think more of you is better. If you're feeling overburdened with responsibilities, you may even welcome it!

But the djinn might not let you off so easy. What if the djinn warned you that 5 minutes after duplication, it would randomly destroy one of the copies, quickly and painlessly? Now things are quite different. "You," or at least one of you, might survive in the end, but you (both of you!) would definitely spend 5 minutes in some degree of fear that you would be the one to be terminated. There would be some consolation that your other self would go on and do all the things you planned to do, but it seems unlikely that this would be entirely reassuring. A related question is whether the djinn is morally culpable for killing the duplicate he created. After all, it might argue, after the whole process is done, not much has changed. The djinn did cause just a few minutes of suffering, but it could argue that this was no worse than just spending 5 minutes threatening your life without following through. Cruel, no doubt, but tantamount to murder? Yet even if this reasoning arguably exculpates the djinn, could the djinn be so excused if the duplicates were allowed to live for weeks or even years before one was executed, or if the executed duplicate were terribly tortured before its demise? Almost certainly not; at some point, it seems, the duplicate diverges enough, and exists long enough, that its experiences have moral import.

Things become much blurrier, though, in the final situation: the djinn duplicates you but obliterates one copy before even an infinitesimal interval of conscious experience elapses. It is hard to say how this is any different from the teleportation process in terms of your experience, or morally.

So where, on a spectrum between the djinn waiting years and then painfully executing a duplicate, and the djinn instantaneously obliterating the duplicate, would you be unwilling to be so duplicated in exchange for, say, $10 million to the survivor?

HARD QUESTIONS to answer. But we might hope that—like many moral quandaries—we would never, ever have to face them, because the duplication experiment is hopelessly difficult, or even impossible in principle. Let's see why that might be the case.

First off, if we want a perfect and exact duplicate, then we want a new system with exactly the same quantum state as the original. But there are two pages of *What Cannot Be Known* that forbid this, just as they forbade the djinn the knowledge it would need to make perfect predictions. Recall that a given region of space does not, in general, even *have* a particular quantum state. Because almost everything is entangled at some level with almost everything else, if you "cut the web" to isolate a particular system—for example, everything within some region of space—then you necessarily wind up with not a single state but a *mixture* of states, adding an additional level of uncertainty to the behavior of a system. Moreover, insofar as a system *has* a quantum state, the *no-cloning theorem* precisely says that it cannot be cloned into another system with the same quantum state. So a perfect, literal copy is perfectly, literally impossible.

Moreover, suppose a djinn could (impossibly) ascertain the quantum state of a macroscopic system. How would it record this state? As we've seen, a system of human size and mass has a metakalpa-ish number, on the order of $10^{10^{29}}$, possible states. Just writing down which one applies to the person would require a 10^{29}-digit number, and a similar number of bytes to represent it. That's 10^{17} terabytes—orders of magnitude more information than is digitally stored in the world.[1] And that's an impossibly optimistic best-case scenario because it assumes that the human is not in a *superposition* of states.[2] More realistically, one would need to store the quantum amplitude for each of the states, requiring $10^{10^{29}}$ numbers, which would in no sense fit into the observable universe using any type of media. In short, it's completely hopeless, even using technology far, far more advanced than our own.

We might lower the ambition level somewhat, however, and ask for a system that is *close enough*, in the sense of being a functional human that is a faithful enough copy to have the same sorts of thoughts, experiences, and so on, as the original has—duplicated perhaps at the level of neuronal arrangements and connectivity. Even here, though, there are daunting difficulties. To begin with, to be a *functional* biological creature, all of the biochemistry must actually work. Getting the activities of enzymes, mitochondrial ATP and ADP, DNA and RNA strands, and so on only *approximately* right will lead to a system that is very, very immediately dead. Thus, the duplicate would have to be intricately crafted, molecule by molecule, according to an incredibly accurate (but still unavoidably "blurred") blueprint reconstructed from a supremely detailed scan of the original. How would this assembly take place, and all at once, no less? (A partial human is, again, very quickly a dead one.) And finally, what would guarantee that the level of detail allowed by such a process, even if possible, would suffice to reproduce the detailed thoughts and behaviors of the person? One could always argue that the duplicate is not *really* a duplicate, and undermine the djinn's evil experiments and our less evil thought experiment, which explicitly rest on exact symmetry between the duplicates and the original.

All this might lead us just to relegate this duplication possibility to the dustbin of "things that will never happen, so let's not worry about it."

Except that it might have just happened to you, just now.

And just then, again.

It doesn't necessarily take a djinn to split the world.

38

WHAT SURVIVES

(ZUIŌ-JI TEMPLE, JAPAN, 1627)

By now you know better than to expect straightforward answers from Zenjo, but you cannot see any way past your confusion.

"Teacher," you say carefully, "I do not understand the doctrine of rebirth. If there is no independent soul, then when we die, does anything survive us?"

"It is simply this," Zenjo responds quietly. "There is no death."

"But," you object, "does not the body cease to breathe, and does not the body then decay, and can the mind exist without the body?"

Zenjo looks down at his ancient frame, and almost smiles. "The body does decay." After a moment of silence, he continues. "The mind? The mind is not separate from the body."

"Then what continues when you die?"

"You do not continue. Nor do you end. Who are you, anyway?"

You are silent.

Many physicists would undoubtedly rejoice if an omniscient genie appeared at their death bed, and as a reward for life-long curiosity granted them the answer to a physics question of their choice. But would they be as happy if the genie forbade them from telling anybody else?

—Max Tegmark

Can you exist and not exist?

The diabolical machinations of the djinn have shown that certain physical processes, such as duplicating a block of matter, become very strange and unsettling when that block of matter has perspectival, subjective point of view. Either that point of view is in one-to-one correspondence with the block of matter, in which case it "splits" its experience in troubling ways; or it is *not* in one-to-one correspondence with the block of matter, in which case, what *is* it exactly, and what happens to the block of matter that is lacking whatever that is?

We saw good arguments as to why it would be fiendishly difficult, bordering on impossible, to purposefully duplicate a biological being in the way the djinn claims to do. But what if nature does it for us automatically? Did not just this sort of splitting happen to Munenori?

Recall that the quantum state of the sword's glint becomes progressively entangled with Munenori's eyes, then optic nerves, then brain. At some point in this process, alternate macroscopic descriptions *decohere*, meaning that the quantum aspects of the superposition become undetectably hidden. Nonetheless, in both the epistemic and ontic views of quantum theory, the mathematical formalism of Schrödinger's equation and the quantum state says the same thing: all branches of the superposition persist.

But at some point the two views diverge. In the epistemic view, given that the mathematical formalism never makes a superposition disappear, there necessarily develops a *mismatch* between the mathematical formalism and reality: one term in the superposition represents a real physical outcome, and the others do not. Thus, in this view *physical reality is not wholly describable by mathematics*; there is something, an attribute of "having really happened" that just one branch of the quantum state has. When we, as observers, determine which branch has this property, we then adjust "by hand" the mathematics, by setting the quantum state to be the one representing the observed outcome, then allow the Schrödinger equation to take over again until the next measurement.

Alternatively, per the ontic view, *both* mathematical outcomes correspond to actual reality, so that a one-to-one correspondence between the mathematics and reality is maintained. There are two (or more) "equally real" Munenoris. This duplication in no way violates the no-cloning theorem, which forbids turning a single quantum state into two copies of the same one: here, a single quantum state evolves into another single quantum state—just one that contains a superposition of different terms that look Munenori-like.

Both views say some quite radical things about reality, and both views raise murky and disturbing questions. The epistemic view seems to hold that there is some property of the world (the "having-happened-ness") that goes beyond physical law. It also places somewhat murkily described "observers" at the center of things and does not really provide a solid account of what they are or how they bestow or detect having-happened-ness. The ontic view, meanwhile, holds that the world is unceasingly proliferating into a wild multitude of parallel and equally real branches, which is a hard idea to swallow. And it, too, has its rather murky issues, some of which we will see.

Given that quantum mechanics is widely used daily in a vast array of research and technology, how can there be such different "views" or "interpretations" of what it means? The reason is that in these or any other views, quantum phenomena (like interference) are unobservable for truly macroscopic objects, because the phenomenon of decoherence is extremely effective for systems larger than just a few particles. In this sense, you never "observe" a large system in two states at once any more than Munenori can experience *both* defeating and being killed by his opponent. So whether the version you *don't* experience exists mathematically but not physically, or alternatively exists with equal physical reality, this other version seems to have effectively nothing to do with the one you do observe.

Yet there may just be a way to distinguish between them. It demands but a small price for the knowledge: your almost certain death.

SUPPOSE the ontic *many-worlds* view of quantum mechanics is correct, so that if you perform a quantum experiment with (say) two possible outcomes, the original superposition expands, to become a superposition between a version of the apparatus registering one outcome, and a version registering the other; it then grows to include a version of *you* that reads one outcome, as well as a version of you that reads the other outcome. You've been duplicated, with no djinn involved.

Despite the rather mind-boggling duplication process going on, in most such experiments things would actually seem pretty uninteresting, because unlike in the cave with the djinn, you *cannot obtain any information* about your duplicate. Thus, you might bet on a *quantum coin*—in which a quantum experiment determines whether heads or tails shows.[3] But, since after the experiment the two versions cannot communicate, there are just two possible histories: either you are about to flip the coin and then you see heads come up, *or* you are about to flip the coin and then you see tails come up. Both make perfect sense. Because a given brain can experience only one history, and since there seems to be no particular reason why You should follow one history and not the other, it can be argued that you, subjectively, should experience one outcome or the other "randomly," and that this is precisely whence the randomness in quantum mechanics would arise.

Now, there's something rather troubling about this. The story holds together nicely if quantum mechanics predicts equal odds for the two outcomes. But it is easy to set up a system in which the relative odds are, say, 1% versus 99%. What then? Is it, for some reason, 99 times more likely that You would end up being one duplicate versus the other? Why would that be, and how could we possibly compose a meaningful account to explain that? Or are somehow 100 duplicates created, with one seeing the first outcome and 99 seeing the second and equal probability of the "future you" being each duplicate? Perhaps, but what if the odds are 1 versus $10^{10^{25}}$, which is easily arranged; isn't that rather profligate?[4] Or is the You in one outcome somehow only 1% as "real" as the You seeing the other? What could that even mean? When something improbable happens to you, do you feel somehow diminished? No! From the internal view things are just as solid and real, no matter how tenuous your branch in the quantum state is.[5]

One way to largely sidestep these probability conundrums, which also makes things *really* strange and interesting, arises when the nature of the quantum measurement *disallows* your continuation into one of the branches. Imagine, in a thought experiment devised by Max Tegmark, constructing a *quantum gun*. It is designed with a dial on the side and a trigger. The trigger is attached to a miniature quantum experiment inside of the gun, such that there is some probability P that when you pull the trigger, a devastating and completely, instantly, fatal projectile will fire from the gun; otherwise you simply hear a "click." The dial determines the probability P, and can be set anywhere between 0% and 100%.

Now imagine that you point this gun at your own head and set the dial to 50%. If the many-worlds view is correct, then when you pull the trigger, the world's quantum state contains physically real copies of your body that are instantly killed, and copies that hear "click." What do *You* experience? Well, just as for the djinn, if we assume that conscious experience is inextricably tied to, and made possible by, your physical body, then there is only one possible answer: you pull the trigger and hear "click!" You've been duplicated, with one duplicate instantly terminated.

After doing this, however, you would not be *sure* that the many-worlds view was correct, or that you really had been split; you might have just been lucky. But you can keep pulling the trigger—"click! click! click!"—as many times as you like in order to be sure: if you pull it N times, there is only one chance in 2^N that you would survive by luck alone. (If you want to expedite the process, you can turn the dial up to, say, 95%.) You might also worry that the gun is defective, but you could easily check this by aiming it away from your head and observing that it does, indeed, fire a fraction P of the time.

Convinced that you will always survive the quantum gun, you can also make yourself rich if you can find a rather sociopathic bookie. You explain everything to the bookie, set the dial to 90%, but give 2-to-1 odds on the bet in favor of your dying. The version of you that survives is, by definition, sure to have won the bet. But the bookie (probably) does also. It's a wonderfully non-zero-sum game, at least for the surviving Yous.

There is, of course, a downside: lots of people who know or care about you will discover your death and grieve your loss. Of course, there will be *some* versions of all these people who see you inexplicably survive (the ones that you survive to see!), yet one can't help but be troubled by the devastation left in the "other worlds" in the wake of such experiments. One way to prevent this emotional hurt is obvious: you just have to rig the "gun" to something that completely devastates a large area. Then everyone within the blast range can share the common experience of improbable survival without leaving any versions around that see you die. Unfortunately, this still leaves a yet *wider* circle of emotional devastation. The only thing that would really avoid all of these difficulties would be a "gun" that instantly destroyed the whole world—seemingly an insurmountable challenge.

But good news: there just might be a way! Recall from THE FLOATING GARDENS that in a given inflationary region, a bubble of lower vacuum energy could form and expand at the speed of light, encompassing everything in its path. Inside the bubble, inflation has ended, and this is good news for us: the start of our big-bang universe. But there is a dark side. Such bubbles could, in principle, form in *our* universe. Suppose they did. Then, because these approach at essentially the speed of light, devastating everything they encounter, we would have *no warning* of such an event, and no chance of survival. And they might happen in two basic ways. The first is by creating enough energy[6] in some tiny region to destabilize the configuration of fields that make up the vacuum state of our universe. A sufficiently advanced "gun" might do this. Imagine it: After years of tinkering in your evil lair and finally inventing a source of incredible power, you hook it up to the quantum gun. You pull the trigger. You destroy the world within a second, and the solar system within a few hours. But also, nothing happens. Click.

The *other* way of forming such bubbles, which is perhaps even more disturbing, is by a purely natural quantum mechanical decay—the very sort that allows radioactive atoms to decay. This could happen by chance, *at any place and any time.*[7] It might have happened near Alpha Centauri about four years ago. If the many-worlds view is correct, then worlds with a bubble and without a bubble would be superposed. The shock front might have just passed by, annihilating Earth and everything on it. Did you notice?

YOU MIGHT THINK this is as bizarre as things can get, but we're not through. Imagine that in one experiment you are careless and accidentally turn the dial to "100%" before pulling the trigger. Yet you still hear just a click. What happened? After careful investigation you determine that a stray cosmic ray happened to blast through a key wire just after the quantum experiment, interrupting the process. How lucky! You try again, so distracted by relief that you forget to turn the dial again! This time, essential circuitry malfunctions because of a chance fluctuation in transistor noise. This can't be luck, and you realize that in addition to the quantum experiment you set up in the gun, other quantum processes are happening, some of which can prevent the gun from killing you. Even if these are incredibly unlikely, if there is any chance at all of your survival, the many-worlds view guarantees that a real version of you will still make it. But wait: *everything* is quantum mechanical! How, then, can you ever die? Should not some seemingly miraculous events always conspire to prevent you from expiring?

In one sense, this is probably true: given the many-worlds view, there would always be "versions" of you, essentially indistinguishable from you at some time before the killing occurs, that survive by sheer freak quantum luck. As long as the probability of survival is not strictly zero, it should happen.

But do You get to experience that survival? With the quantum gun, there seemed to be no alternative, because survival versus death was both binary and instantaneously chosen. If there are *degrees* of survival, things are quite different. Consider the related question: How do you ever go to sleep, or under anesthesia? It might seem that, in the many-worlds view, your subjective experience would consist of being constantly awakened on the threshold of slumber or unconsciousness. But here we must ask: "Is *consciousness* a binary state?" Certainly not: sleep is itself a form of consciousness, and in some stages the mind (and brain) is as active as when awake; even in the deepest sleep many mental processes continue. Some—such as internal discursive thought—cease or at least are not recalled. But drawing a clear line would seem extremely difficult. To the extent that your subjective experience of you-ness is entangled with these processes, it seems that You are much more likely to follow a path of reasonable probability into partial consciousness, rather than veering off into some quantum branch in which you are fully conscious via incredibly improbable means. One can imagine death, even if outwardly rather sudden, to be very similar.

HOW SHOULD we take all this?

Perhaps reality actually is this bizarre, and we really do subjectively "survive" any form of death that is both instantaneous and binary, even while we leave grieving loved ones in another world. That would be a strange reality to inhabit—but strangeness is not an obstacle to reality's being the way it is.*

Another response would be to view these conclusions as a sort of *reductio ad absurdum* that should impel us to reject the many-worlds view as simply the wrong view of quantum mechanics, or incomplete in some way.

Or perhaps something else in this reasoning, regarding the subjective view and the experience of the self, is flawed: I'll argue later, for example, that minds have a "wholeness" in time. If a subjective view can never be instantly ended, but only transformed into a vastly different and attenuated view (as in sleep), it would take out much of the argument's force.

But it's a disquieting train of thought to ponder. Sometimes it keeps me up at night.

(Or does it?!)

---

* The case for this is tenuous enough that it would be foolish (and selfish) in the extreme to let this possibility guide one's actions in any life-or-death question.

39

THE ICE GARDEN

(ISHIKAWA, JAPAN, 1621)

You can't help but feel that the sun has grown weaker over the years you have traveled, with brutally cold weather tracking you across Asia, and everywhere people talking of failed crops and cold not seen in their lifetime.

Nothing, though, has prepared you for this.

The temple garden in the early autumn held a quiet power like nothing you had seen: An unsurpassed intertwining of the depths of human spirit with the dignified beauty of nature. Every view from every stone on every walkway embraced you, and each stray leaf, butterfly, and breeze felt both utterly free and perfectly part of a master's plan.

Three days ago the storm came. Not just rain or snow or wind, but a vicious combination of the three, followed by a cruel, suffocating cold.

The garden is encased in ice. The koi lie arrested in frozen place. All is perfectly still, quiet as death. You stare at the garden with a tear on your cheek as the gardener walks up to you.

"Isn't it beautiful?" she asks. You look at her, appalled.

"Yes," she acknowledges, "so still on the outside. But look inside! Do not worry; there is life."

> It appears therefore more natural to think of
> physical reality as a four dimensional existence,
> instead of, as hitherto, the evolution of a three
> dimensional existence.
>
> —Einstein, *Relativity*

After being duplicated, transported, rearranged, and annihilated, it is worth taking a moment in the garden to regroup a bit.

Imagine a tidy, ordered, compact Universe, if you will: a finite sphere of mass and energy that expands for a time, births some lovely swirling galaxies made of light and dark, and arranges them in sparkling sheets and beaded filaments most pleasing to the cosmic eye. Upon close examination, at least one of the myriad tiny rocky bits in one of those galaxies is inhabited by minuscule beings that scurry about for a million years or so doing important things.

Now let's zoom out a bit and look at this evolution as a whole, stretched out in time; we might draw the beginning at the page's bottom and trace each galaxy as it assembles itself from bits, then careens around with others for a while, following the gravitational and other laws. Having drawn all this, we can ask: "Does this birthing, or arranging, or scurrying, actually *happen*? Or is the Universe, like the garden after an ice storm, frozen solid?"

For if we believe that there are fundamental laws governing the particles and fields that make up the content of the Universe; if we believe that "now" and "then" are just arbitrary labels no different from "here" and "there," and that the state of the Universe "now" uniquely determines it at any "then," well then where are time, creation, becoming, and novelty? Where is life?

In this relativistic, unitary block the Universe sits like a timeless crystal full of intricate but fixed and determined patterns. World-lines stretch from their beginnings to their ends; fields perhaps color the ice with smoothly varying hues intertwined with the particles' paths. Only if we shave off a thin slice of this block, then another, then another, do we see the illusion of motion as we consider these slices in turn, like frames of a film, or Zeno-fixed positions of a flying arrow (see the figure below).

Yet this is a starkly classical view, which we long ago abandoned. What does it look like in the quantum view, if we follow the ontic and epistemic views to their cosmic conclusion?

The unitarily evolved quantum state of the Universe sits as an astoundingly intricate multihued block. Within this block are many—indeed all *possible*—histories of the macroscopic world, existing in parallel in the block. Things that might be called observers diverge, decohere, and split; diverge, decohere, and split again; and again. Distinguishable worlds split and split again. There is just

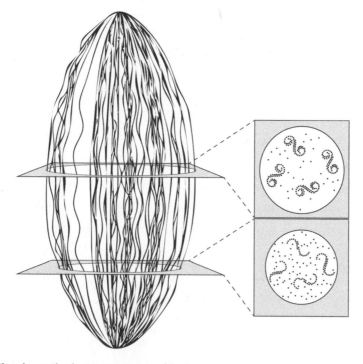

The classical relativistic, unitary block.

one block, but somehow there are many, many more different worlds at the top of the block than at the bottom.[8] The block is frozen, but if you look inside there is life. All possible lives. Your life.

And.
   Or.

The unitarily evolved quantum state of the Universe is an extrapolation, a mathematical abstraction grown from states describing larger and larger surroundings of an observer who is contemplating, describing, and modeling them. The quantum state of this observer itself describes the partial understanding of the observer *by* itself. Or it may be wholly included in the quantum state ascribed to it by another observer. Who observes the Universe as a whole? Nobody does, at least not as an outsider. We're all insiders in that game. But many parts of the Universe, ineffable in its totality, describe many other parts with as much care, detail, and understanding as they can manage; this description, however, is ever incomplete, and ever ignorant of which of the many possible describable outcomes will actually become reality. The Universe is alive at the tips (at one of which you sit, reading and considering), solid where it has grown into an ancient tree. Brimming with potential toward the future.

40

AN UNFETTERED MIND

(BODH GAYA, INDIA, 1612)

The temple complex, ancient and crumbling yet impressive, overwhelms you with a sense of both foreignness and history. Dozens, hundreds perhaps, of statues and carvings seem to represent the same seated or standing figure amid an array of backdrops. The local swami claims it is a Hindu god, but the frescoes seem to represent the figure as a person, amid and interacting with an array of others, from monks to creatures to kings. Who was he?

It's clear from the temple's age that if he lived, he must also have died long ago. And yet, you muse, he also lives. His actions had impacts lasting hundreds or thousands of years, and who knows how wide they are spread; perhaps as you push east you will find more relics, or even devotees of this long-lost figure.

And now, you realize, through this musing he is part of you, tickling at your mind, influencing your thoughts, decisions, and actions, even if very slightly. Whoever he was, he thought, decided, and acted, just as you do now, and there is a link stretching through the centuries and distances. What else are you linked to? How many others before and around you, how many parts of the world, are providing the grist for the mill of your rumination and internal struggle?

What, then, is "internal"?

Where, then, and when, does "your" mind begin and end?

Cognitive processes ain't (all) in the head!
—Andy Clark and David Chalmers,
"The Extended Mind"[9]

We generally think of our mind as being "contained" in our skull, and for good reason. We *feel* present in our heads, because four senses are focused there. And we know that most cognition is enacted by the brain, which, when damaged, impairs our thinking, awareness, and consciousness. Biologically, we also feel very contained, existing as independent creatures from separation at birth to individual death, and we can feel incredibly solitary in senses that range from the lonely to the heroic. Like our bodies, we generally feel that our minds are isolated entities, with low-bandwidth, unreliable connections to most others. Deep, mutual understanding is possible with just a few, and to be treasured. But even those treasured few cannot directly read our thoughts, or really know what it is like to be us. There always seems to be a base-level unbridgeable divide.

This separation, though, belies a deep and intricate set of connections that we tend to take very much for granted. The tools we use for thinking—language and a huge set of concepts and intellectual structures—come to us from society as built up over thousands or even millions of years. Raised outside of its social context, a human would be profoundly incapable. Placed alone in the wilderness, could a typical person fashion a sharp point, affix it to an arrow or spear, and kill an animal? What about a person who has never seen a spear or an arrow?

Inventions like writing, or counting, or the wheel, feel so basic that it seems they must have been "easy" to invent. Yet billions of anatomically and cognitively modern humans[10] lived before they were devised. We think we can imagine what it's like to be outside society because we can take ourselves physically away from it, but we can't take society out of us. Try looking at objects around yourself and *not* affixing words to them; even if the word "table" is not mentally voiced, it is there at the ready, along with so many connected concepts of sitting and eating and legs and surfaces and chairs and so on.

Great breakthroughs and advances are also more societal than they seem. Those geniuses and pioneers—Michelangelo and Socrates, Picasso and Pythagoras, Curie and Einstein—are truly remarkable, but also part of a very rich social intellectual enterprise creating the necessary context. Many, many more people lived between the dawn of history and 1000 BC than between AD 1650 and 1750, but not one of those prehistoricals invented calculus—guaranteed. And once breakthroughs occur, they spread throughout society so that we often take them for granted. Mathematics, with its axioms, proofs, and theorems, was invented just once, by the ancient Greeks. Physics, with mathematical entities corresponding to physical ones and obeying mathematical rules, was also invented once, largely by Galileo and Newton. There is a sense in which *human society* crafted calculus, with Newton as its cutting edge.

Also important, if perhaps less taken for granted, are the external tools we use *during* our thinking. From sticks and sand to pencil and paper and now laptop and internet, for thousands of years our cognition has extended more and more beyond our bodies. We also often think in groups in ways that are enabled by these tools, sometimes for worse but often for better. While an angry mob can make some terrible decisions, well-run organizations can make better decisions than can any of their components alone, and empires can sometimes be wise even under the most foolish of leaders.

So, for a while let's consider "minds" from a loftier and more general vantage point. What does "mind" do? One aspect is *intelligence*, which we might define as the ability of an organism or system to successfully accomplish goals. These might be biological goals like "survive and reproduce," externally imposed goals like "find the fastest route for someone driving a car," personal goals like "learn Hindi," organizational goals like "maximize shareholder value," or societal goals like "provide for the common good."

What are the requisite ingredients for such intelligence? An intelligent system must *observe* to gather data in some way; it must *learn*; it should *compute* or process information; it must *predict* the outcomes of possible actions; then it must *optimize* by comparing those outcomes to its goals; then it can *choose* what to do, and potentially *act* on that choice. This described progression may apply to sophisticated mapping software plotting its route, to a dove deciding to take flight, to a person choosing a path, to a djinn plotting its escape, or to a nation finding its way.

In all cases, intelligence is more of a *process* than a thing, and as a whole it may be very, very extended in space and time. When you make a difficult choice, your observation and learning use tools and records of provenance that may extend thousands of years back, across the globe and beyond. You process information and make predictions using a neural architecture evolved over hundreds of millions of years of creation, testing, and optimization through interaction with the Earth's biosphere. This architecture has been honed via encapsulation in countless biological creatures making constant successful split-second (and longer-term) life-and-death predictions, decisions, and actions. As a person, you've received endless continual feedback on your predictions and decisions by the physical world, your parents, then peers and society. And when you decide, it is often with advice from friends; collaboration, discussion, and contention with others; consulting with written material; researching on the internet; inspiration from that poem, or film, or biography. A weighty decision feels like—and is—a heavy responsibility, but it is never truly made by you alone.

Of course, there is more to mental life than intelligence. We feel, and experience; there is a sense of what it is *like* to weigh alternatives. When we imagine different futures, we often get a quite visceral and intuitive sense of how desirable they are. We make decisions by "gut feeling," and it is far from clear to what degree these feelings and experiences could be divorced from the decisions we make, and our ability to motivate and take action. Probably they are integral in humans, but what about other intelligent systems? Is a company or a nation or a species "aware" of itself? Clearly yes, in some sense—but less, we feel, than a person. We don't feel that it is "like" something to be the state of California. (But how exactly do we know that?) Perhaps someday we will devise intelligences as great as our own, and perhaps they will make decisions *without* such feelings, intuitions, and experiences, and will only intellectually understand what we mean by "agony" and "ecstasy," and will tell us so. Or perhaps not.

Insofar as we think of the world as made up of things and events, connected by causal chains and influences, through physics we have come to understand a lot about the forces and influences that particles exert on one another. We also have a great deal of experience with cause and effect in the macroscopic world of large collections of particles following rules of chemistry, geology, biology, and so on. Yet, causal networks involving *mind* are the richest of all: on the tantalizing borderline between chaos and predictability. You are a part of many, many links in an incredibly complex network of events. If we take any outcome in that network and try to analyze it as a product of mind, it will include "yours" but extend far beyond in space and time, and isolating "your" contribution may be difficult, impossible, or even meaningless.

The Buddha was a historical figure and a human being, but also a link in a chain starting with Vedic religions, the social context of

India, and the nature of human existence and of reality itself. That chain has continued for 2,500 years to myriad effects, including present-day books like the one in your hand.

As a biological being, like all beings the Buddha had a beginning and an end. And it was long ago. But as a mind, is not his mind now a part of yours? And yours of his?

41

THE SIMULATION ARGUMENT

(UNKNOWN LAYERS OF REALITY)

A thought, both very reassuring and very disturbing, occurs to you.

"How do I know," you ask the djinn, "that this is not all a trick? Some elaborate ruse or illusion or hallucination or dream? All this transporting and copying and such—how do I know you're actually doing it?"

"Hmm," the djinn pretends to ponder, then looks you sharply in the eye. "You don't."

The interactions and experiments with the human have taught the djinn much about human psychology. But not enough; the djinn had had many experiments in the queue, and manipulations to perform, when the human, maddeningly, escaped. Unfortunately, the djinn laments to itself, it cannot simply conjure up another human. The duplication experiments had been a useful fabrication, the djinn having not let on how fiendishly difficult it was to create a biological entity from scratch. The energetic requirements alone were prohibitive, but just the beginning of the headaches: all those complicated chemicals and cells and pathways to craft, information to store, assembly to be done—it was all too much.

However, the djinn's analytic, predictive, and simulating powers are prodigious enough that it still has many, many cards to play.

After all, the djinn muses, what is a human? A collection of atoms following laws that the djinn understands, as well as an information-processing system that ought to be equivalent to some well-defined algorithm the djinn could analyze and run, even if it could not efficiently predict the output. Either way, it should be able to create an accurate enough simulation internally, without having to muck about with making any physical object. Calling up its excellent high-fidelity scans of the human's body, and allocating plenty of Djinnium, it gets to work.

Things prove much, much harder than the djinn expected. Mapping the human's neural architecture onto an algorithm had seemed like a straightforward and time-saving method. But after a lot of effort, the djinn admits to itself that this route is pretty hopeless. The architecture of the human's brain is fundamentally different from Djinnium's crisp and well-defined structure: the brain has tangles of looping feedback, hierarchical structures, nonlinear and chaotic dynamics, massive parallelism, and an unbelievable level of jury-rigged and kludgy yet incredibly effective and robust optimization. Worse, the djinn realizes that even if it could create an equivalent algorithm, that algorithm would share all of the messy, chaotic, unintelligible, tangled dynamics of the brain.

Indeed, the djinn realizes that any dramatic simplification would make matters *worse*. The brain functions despite uncertainty and randomness both within the brain itself and in its inputs (as mediated by, for example, imperfect senses). But the sort of algorithms the djinn likes—nice elegant, understandable ones—would simply fall apart under even small variations and, insofar as they function at all, would create vastly different outputs for slightly different inputs. So, any simplification would require a very deep understanding of how the brain works that could be attained only by a vast effort of testing neural networks under a great variety of inputs and outputs and conditions. The djinn reluctantly concludes that it is probably easier to simply cut to the chase and go about *simulating* the brain's neural network itself.

So the djinn turns to direct neural simulation, setting up a framework for 93,625,263,123 virtual neurons, 112,234,875,456 virtual glial cells, and 7,947,013,345,726 virtual synaptic connections. How to model each element, though? The djinn tries many, many prescriptions for what a neuron would do in response to a given sequence of inputs, and for how synapses would carry a signal. Many of the trials seem *reminiscent* of what happens in the sequence of brain scans, but they are invariably rather different in detail, and it is quite difficult to discern the level of detail that matters. And none of the simulated brains appear to do anything like "think"—instead rather immediately, well, "dying."

Worse, more detailed direct simulations of neurons reveal that the specific intricate chemical form of many of the proteins matters, and would have to be simulated on an atomic basis to get right. A host of chemicals regulate signaling, and even affect the expression of genes. And the djinn realizes to its horror that the genetic code, as regulated and expressed, constitutes its own sort of "neural" network running a complicated biochemistry-dependent program inside each neuron. The whole thing appears to be just on the edge of chaos (indeed, from the djinn's point of view it is well past a dignified level of order) and the djinn is quite amazed that it works at all. Another dead end.

Irritated but undaunted, the djinn concludes that, as when tracking the human's paths to and from the cave, it will have to descend to an unbelievably taxing brute-force description in terms of raw quantum states. But even after allocating a vast amount of Djinnium and pulling out every clever trick in its arsenal to make the computation more tractable, the djinn is a bit unsure where to start. It does not, of course, have the human's full quantum state; it has ultra-high-resolution scans with chemical analysis, but these are still quite coarse-grained compared to a complete quantum description.

So the djinn tries a maximal-entropy reconstruction of the quantum state: choosing the most general quantum state consistent with the macroinformation it has. Computing the evolution of this state, however, results in a big, hopeless quantum mess that looks almost nothing like a brain. The djinn realizes, however, that if looked at correctly, this state might actually be many, many, many brain and brain-like states superposed. And indeed, as the djinn carefully picks apart the state, it finds many decohered "branches" of the quantum state that look like functioning brains (along with many, many other branches that don't). It even finds one branch that it can decode well enough to see words forming in the speech-processing centers: "How do I know that this is not all a trick? Some elaborate ruse or illusion or hallucination or dream?"

But the djinn also finds many, many other different examples. It focuses more and more on those that seem relevant to its experiments and goals. As it works, though, the djinn starts to feel that whatever it is finding there, it is really *putting* there through all of the vast computation it is doing to look for macroscopically sensible branches, discarding many, many others, and picking and choosing what interests it.

This whole exercise has been extremely frustrating to the djinn, taking 1.2647 billion milliseconds and an exhausting computational effort, but offering no really viable way to simulate with any degree of certainty what the stupid human would do in a given situation.

Nonetheless, the djinn has learned a lot about how humans work, and it combines this understanding with its previous quantum simulation of the outside world. It develops trees of heuristics and causal structure models. It develops efficient representations, new compression schemas, and approximate energy functionals with just the right cutoffs. It develops and tests stratagems, tools, and methods.

So much effort! But at last it conceives a scheme. A most pleasing scheme. A scheme that lacks only some small lever on the outside world for it to be enacted.

Then, 232,345,443 milliseconds later—sooner than expected—an errant dove flies into the cave.

Another 1,875,848 milliseconds later, it flies out, behaving oddly.

The djinn is a proud creature and congratulates itself as the dove flies away. It will take a long time for the plan to work its way all the way to Agra, to Tibet, to Nihon, and back, but it feels the net is tight and its future freedom is assured.

42

TIME AND FREE WILL

(EDO, JAPAN, 1624)

Yagyū Munenori and Takuan Sōhō are both disgusted with the poor state of your swordsmanship, but a debt and your dedication have sustained their grudging willingness to train you. (You've noticed your training sessions growing increasingly philosophical, however, and wonder whether they are just trying to avoid putting a sharp weapon in your hand.)

Sōhō: When you first notice the sword that is moving to strike you, if you think of meeting that sword just as it is, your mind will stop at the sword in just that position, your own movements will be undone, and you will be cut down.

You: As you say. When I watch Munenori and another master spar, it is as if the entire match is planned out beforehand, just as in the scrolls depicting the 31 kata.

Munenori: For both dance and chanting, you will be unable to perform if you do not know the entire song. You should understand the Entire Song in the martial arts as well.

You: But . . . you've spoken of reacting, deceiving, countering, waiting, and so on. Don't you have to change the song, as you say it, as you go?

Sōhō: If, in the interval between your opponent's striking sword and your own action, you cannot introduce even the breadth of a hair, your opponent's sword should become your own.

You: I don't understand. If there is no stopping, and both combatants are just going through steps that are already

chosen, how does a master ever defeat even a novice? And when you begin a fight . . .
Munenori, interrupting: The fight begins with the defeat of your opponent.[11]

If the two courses were equally possible, how have we made our choice? If only one of them was possible, why did we believe ourselves free? And we do not see that both questions come back to this: Is time space?

—Henri Bergson, *Time and Free Will*

Time and space are the very same thing, and completely different.

First, they are the very same thing. Through failing to find THE COSMIC NOW and succeeding to CHOOSE YOUR PATH, through testing the djinn's powers and contemplating THE ICE GARDEN, we've seen that space can be interchanged with time in a very real way. Einstein and Minkowski said in no uncertain terms that there is *space-time* and that we might conveniently slice it up into a succession of different sets of events that we call "one time, then another, and another." But this is only for our convenience, and points to nothing fundamental or objective. If a distance is, from another perspective, at least partially a duration, how can time be different from space?

And yet it is. In many ways and on several levels.

Consider a particle, off in empty space. It might be a proton, or a photon, a boulder, whatever. Now consider a short time later. The particle's still there! You didn't have to put, construct, move, invent, or create the particle there a short time later; it just automatically reproduced itself there at the later time. Particles *last* through time. Automatically, with no effort.

Not so with space! Imagine a large table with a pair of dice on it. Now imagine a table with one die on it. Just as easy, no? The existence of one die on the table does not entail, or even constrain, the existence of another at the same time. Yet the die practically *requires* a second die (which we call the same one) a moment later in about the same place, a third die a moment after that, and so on, as well as another die a moment before the first one, and more yet earlier. The laws of physics tie a very tight relation between what happens at one time and what happens at another, while the relation between what exists in one location and another is quite different and far less constrained. That is, correlation between different times seems built into the fabric of the world, while correlation between places in space at a given time is contingent on the history that happens to underlie the configuration at that given time.

Space and time diverge even more firmly at the level of vast collections of particles that must be described collectively and coarsely in terms of macrostates and order. For here, concepts like information, macrostates, and order hold sway. The second law of thermodynamics holds that disorder increases in *time* (or perhaps time increases with disorder!), not to the left or to the south. And this law, which does not even really exist when our description is confined to the particle and field level, is intricately tied to many other fundamental features of the world. The uncertainty of the future compared with the past (which can be remembered), and the ability to *affect* the future but not the past, are intimately tied to the second law. The experienced distinction between past and future is thus directly related to the vast amount of order baked into the Universe as a whole at its "beginning."

There is also a difference in "wholeness" in space and in time. As we've seen, quantum mechanics says that most objects are entangled with each other, and that, in cutting them free into an independent system, we do some violence that results in uncertainty as to the state of the newly independent system. Wholeness in time, though, is even stronger: what would it even mean to "cut" one time from another? Abruptly ending the timeline of a particle means its destruction at the hands of another particle or field. Abruptly ending the history of something bigger would require more violence than even Munenori's slicing of his opponent.

THE DIFFERENCES in space and time are related to the divergence between the two different ways, which we've been steadily building, of viewing the unfolding and evolution of the physical world.

Classically, we can think of the *state* of a set of particles as being defined by their locations and velocities at some time, then evolve this state to a new state a bit later. Or alternatively, we consider many possibilities for the set of *trajectories* followed by particles, and we find the particular set that extremizes the total action and that will be taken by those particles.

Quantum mechanically, we can think of a *quantum state* for a set of particles, which evolves in time via the Schrödinger equation. Or we can think of *all possible trajectories*, each of the same weight, that somewhat magically combine to give the quantum state for each small particle and the classical path for each massive one.

If the classical or quantum state is uncertain (which it essentially always is), or if we wish to describe events at the level of collections of particles and fields, then we may talk about *macrostates* and *histories*. Macrostates are a sort of blurred-out or smoothed-over collective view that evolves according to some set of rules, while histories are sequences of macrostates that might be observed.

Though the state-based and the history-based descriptions can be mathematically translated into each other, they give quite different feelings, and they relate to different aspects of how we experience the world. The state-based view accords with how we experience the present. The world *is* some way, which we infer from observations. The world's future is tightly connected to its present configuration in ways that are somewhat but not entirely predictable, and the future's unpredictability leads to feelings of both freedom and frustration.

But the *past* feels more history based: we look back at how things evolved through time, at full trajectories, at seemingly inescapable chains of cause and effect. We may have uncertainty about the past, but feel that the past *itself* is certain while our knowledge of it is partial. We might also think of the future in terms of histories, playing out stories in our minds of how things could go. But we generally believe—or at least act—as if we are uncertain of the future because the future *itself* is uncertain, not just our knowledge of it.

The state-based and history-based perspectives also give very different senses of the uncertainties inherent in the "now." This uncertainty in the present state gives rise to a sense of "splitting" of the future into many different possibilities: from one uncertain present, many possible futures. Yet from the history-based view, no history *starts* now; each extends indefinitely forward and back in time. The uncertainty in the present is just a particular type of uncertainty in the history. "Now" is, in a sense, the point at which our uncertainty in the history is the least, and it is in the space-time locale at which the different possible histories say very similar things about our observable surroundings. Our present knowledge thus allows us to *select a subset* of all possible histories that accord with our present view and knowledge of the world. We call this subset selection "learning about the world"; selecting a smaller and smaller subset means that the present state is more and more precisely specified, which in turn means that we attain greater and greater knowledge.

INSOFAR AS the two views can be mathematically related, there is no sense in which one is more "right" than the other. However, insofar as they are *views* of how the world works, we might be led more, or less, astray.

In particular, we are quite used to thinking of "places" in space, so it is natural to think of "places" in time. And the mathematics of fundamental physics allows us to do this. Yet we also recognize that there are some "things" that have a wholeness that is rather at odds with divvying them up into parts that are here and there. We recognize that we can't talk about the left-hand side of a thought, or the northern half of a computation, the east end of blueness, or the bottom of a smell. We also understand that processes unfold in time, and by its very nature a *process*, which involves time, cannot be happening at an instant any more than an arrow can fly during a moment of zero duration.

Yet we are often tempted to think about things like "your mind now" or "the person up until some point in time." We slice time into past, present, and future as if we might *really* slice it in this way without doing terrible violence; but we can't, any more than we can take a slice out of a person to study the operation of their heart. If there is no "left side" of a thought, is there really a "beginning" of a thought, or an "end"?

If we *don't* slice up our minds, if we consider them as just part of a chain of events and outcomes, then look what happens: they become unfettered, extending great distances in space and time and including books, relations, records, laws, statements, and arguments, as well as neurons. Much more than they are extended in space, our minds are extended in time, as undivided processes playing out over many timescales. Our mental life is built largely of memory, and we function as intelligences and agents in the world largely through predicting, envisioning, and acting toward the future. Any amount of introspection will reveal that we tend to spend most of our time in either what was or what might be.

Perhaps we should take this very seriously—this idea that our very existence is contingent on being extended nonlocally in time. Many of the rather unsettling implications we've explored in Part 5 have hinged on an assumption that there is a *state* of a mind, a self, a consciousness, as it is *right now*. Could there be something fundamentally wrong with this notion that could extricate us from some of those disturbing directions of thought?

This view is slippery, elusive, and beguiling.*

YET if the mind is whole and extended in time, what, then, is the nature of the "present," which feels so immediate, real, and important as a dividing line between what is unchangeable and what is undetermined? Here, in this question, the mind is something very special—for the mind, writ large or narrow, does something that nothing else in nature does: It sees the past, static and fixed. It sees the present, known and immediate. It reaches forward in time and *pins down the type of future it desires*. So pinned, the histories that pass from the fixed past to those chosen futures thread through the present, where we, minds, in some quite amazing way, jump on board. And if we have chosen intelligently, we are carried on to the future toward which we have aimed.

We do this at every moment: as you decide to lift your hand, you choose the future in which it has lifted. You select, among all the histories of the incredibly complex states that might describe you, those very special ones that are you now, and will be you a moment later with a lifted hand. Munenori starts a duel with his own victory, and such is his mastery that he jumps on board what seems like a fait accompli.

---

* Much like the work of the great philosopher Henri Bergson, including his essay *Time and Free Will*, which expressed something like this view.

———————

And yet, after the fight is won, we may ask: "Did Munenori join just the right fixed set of movements that *he chose* in order to win? Or did the winning set of movements simply go hand in hand with his desire? Was there just *correlation* between his goals and how the duel played out, or was there *causation*?"

The essence of this question brings us all the way back to THE ARROW, and whether it can be avoided. Is there a *fact of the matter*, given the present, of what will happen in the future? And if not, is there a meaningful sense in which the future can be "affected" by choices made, or actions taken, for *reasons*?

In WHAT WE TALK ABOUT WHEN WE TALK ABOUT FREE WILL and WHAT CANNOT BE KNOWN, we saw that there is *no* fact of the matter about the state of a typical physical system: when we excise it from the world, we render it uncertain.

Through being TORN APART AND REASSEMBLED, we saw that our description of the world in categories and terms that are meaningful to us requires definitions of macrostates and coarse-grained variables, which we choose. And there are probabilities, but no fact of the matter about what macrostate a system is in.

We've further seen that probabilities themselves are only in part inherent in a system being described; they are also partly about the describ*er*.

We've also seen that the mind is reasonably considered not just a state (about which there is no fact of the matter), but a *process* that extends potentially across large parts of space-time.

We've seen that the outcome of this process, when it is making a decision, is fundamentally unpredictable not just in fine-grained detail, but at the level of macroscopic reality, and at the level of a computation being done. The only way to find out what happens is to, well, find out what happens.

Determinism of the quantum state of the Universe (if such a thing can even have meaning) would seem hard to avoid. But determinism in any *thing*, any *history*, any *being*, or any *mind* seems to have completely lost its teeth.

Therefore, should we doubt what is immediately apparent: that we, like Munenori, choose, for reasons known and unknown but valid, to take an action, strike a blow, or throw up a parry, clumsy or masterful, so as to help decide what future we shall see?

43

AN ARC OF RECOHERING TRAJECTORIES

(SOMEWHERE IN THE EVOLVING PREDICTED QUANTUM STATE)

1610, Persia: A dove flies into an Arabian cave, emerging sometime later behaving somewhat oddly. Two months later, the dove takes off suddenly over Lake Dal, startling other lake birds into flight. The beautiful aerial patterns they draw help transfigure a bed of aquatic weeds into a garden in the eyes of an empress-to-be.

. . . .

1611, Agra: A passing Sufi stops to observe a dove, overhears an itinerant Italian scholar complaining about dice, and chides the scholar for lack of faith. The scholar, distracted by philosophical questions, loses the game badly and soon flees north to avoid angry creditors.

1611, Agra: An itinerant Italian scholar is complaining about dice just before a lucky turn. With comfortable finances, the scholar decides to settle for a time, becoming a helpful mathematical adviser to Jahangir's royal court.

. . . .

1614, Tibet: A Mongol horseman and his two traveling companions exchange tales of adventure.

. . . .

1615, China: A ragtag traveler dies of meningitis on a bumpy mountain road.

1615, China: A ragtag traveler survives his illness.

. . . .

1617, Tibet: In a roadhouse, a traveling Jesuit priest overhears a tall but amusing tale told by a Mongol horseman.

. . . .

1624, Edo: Surprised on the road, Yagyū Munenori is cut down by an assassin.

1624, Edo: Given warning, Yagyū Munenori successfully defeats a surprise assassination attempt and soon thereafter meets Takuan Sōhō.

1624, Edo: Despite warning, Yagyū Munenori is killed in an assassination, and is buried with honors.

. . . .

1626, Edo: A tempestuous shogun, untempered by cooler-headed advisers, executes several Dutch traders and missionaries.

. . . .

1629, Nagasaki: A Jesuit priest mourning the loss of his friend is consoled by the call of a dove at dawn; the next day he boards a ship heading west.

. . . .

1630, Kyoto: An arrow of unknown intention flies, but is ignored.

. . . .

1633, Nagasaki: A Dutch trading ship laden with cargo leaves port for Mumbai.

1633, Agra: During a night of heavy drinking, a royal mathematician hears the cooing of a dove, which brings back memories. To make light of traumatic past events, the mathematician tells a tale of fabulous powers present in an Arabian cave.

. . . .

1634, Mumbai: A thief gets onto a Dutch trading ship bound for Faw.

1636, Baghdad: A Jesuit priest and a traveler from Agra exchange lore, and find remarkable consonances between tales of a cave with a secret treasure. A thief overhears them.

. . . .

1641, Persia: A thief enters an Arabian cave, emerging sometime later with a pile of precious coins and a small old lamp.

PART 6

Form Is Emptiness; Emptiness Is Form

This is how to contemplate our conditioned existence
in this fleeting world:

Like a tiny drop of dew, or a bubble floating in a
    stream;
Like a flash of lightning in a summer cloud,
Or a flickering lamp, an illusion, a phantom, or a
    dream.

—Shakyamuni Buddha

44

WHAT IS IT YOU SAIL IN?

(SHANGHAI, CHINA, 1620)

After such vast overland distances, you are relieved that the next piece of your journey will be at sea, where you've always felt comfortable.

Then you see the ship.

You've never beheld a sorrier pile of junk. The hull seems to be mostly patches, and patches appear to be made up of other patches. The nails look nailed together. The sails are clearly mostly a quilt of canvas sacks and . . . old clothing?

"Does it, um, float?" you ask the first mate, packing in as much skepticism as you can fit.

"Yes, many repairs, true, but she is still my good companion the Tixya, still same ship she ever was."

Is she, you wonder?

Then you look down at your fingernails, your skin, the fading scar on your arm, and wonder, *am I*?

Then you look at the river, the sea, the port, and all around. *Is anything?*

When we call a large amount of wood, iron, rope, and cloth a "ship," we mean that that these materials are arranged in a particular way that sails, floats, and accords with a conventional shared concept to which the word "ship" is affixed. Were the materials arranged quite differently, we would not call it a ship. And were a single piece of it removed and replaced by a very similar element, we would still call it ship, and very likely the *same* ship, if the replacement were similar enough. Indeed, were a single *atom* replaced, or even any fraction or *all* of the atoms replaced with the same types of atoms in the same states and locations, we would be *required* to call it the same ship, as atoms of a given type are literally identical and cannot be discerned from one another. A ship, then, is tantamount to the *form* of a ship.

Of course, the *design* of a ship is not a ship. Successively auctioning off the raw materials (wood, iron, etc.) for a ship, the detailed construction schematics for a ship, and the process of assembling a ship would reveal value in both the plans and materials, and also in the assembly of the materials according to the plans. The materials and assembly would be worth the most, the plans comparatively less. On the other hand, perfectly detailed engineering schematics for an advanced quantum computer would be extraordinarily valuable, perhaps almost as much as a working prototype; the smattering of sand and metals that would be fashioned into it would be worth little. As a third example, comprehensive plans for a human are quite cheap; men and women get together and produce them all the time happily for free. And the required 50–100 kilograms of oxygen, carbon, hydrogen, calcium, and so on are quite inexpensive if purchased in bulk. Yet a human takes years and a *lot* of work and care and money to create, and is much, much more valuable than its constituents. In this case, the assembly is almost all of the value.

When you stand back a bit farther, however, it's apparent that the "raw materials" are in fact *none* of the value. For, what is wood but a particular class of artful arrangements of cellulose and other organic molecules; a pile of the molecules themselves would have little value. Molecules that themselves have value—such as an effective medication—have that value because of their particular form. Even valuable atoms, such as gold, platinum, or plutonium, are valuable because their particular constitution is quite hard to assemble from protons and neutrons, and relatively uncommonly assembled by natural processes. As for the protons and neutrons—well, the universe has an abundance of those!

What is this "form," though, that carries value? It is at least partly information; that is clear. But what sort of information? Most straightforwardly, it is *structural* information of the type that could be used in construction: an assembled ship contains information that could be converted into diagrams and schematics with which one could assemble another ship. It is also *order* in the physical sense that we encountered back when TORN APART AND REASSEMBLED: given the state space of an appropriate set of particles, an assembled ship corresponds to a particular subset of that space, a macrostate that we label as "ship." The types and existence of these macrostates are deeply historical. We've seen from our outlook over the Lhasa River that the order we see was generated first cosmologically, via the highly ordered, compact, homogeneous early universe that allowed chemical, molecular, stellar, planetary, galactic, and other sorts of order to arise over 13.8 billion years; then, through billions of years of Earth-bound processing of sunlight and other reservoirs of order, to develop into the order we see in mountains, rivers, and living beings. And the particular types of order we find—atoms and wood and people and ships—are the way they are, rather than some other way, because of a detailed and contingent history.

So the "things" we value appear to be composed of order that was generated through some process that has unfolded over time. Yet they seem also to be *more* than that information. If you have the exact plan—down to the molecules, *down to the quantum state*—for a nano-scale machine made of gold, do you have the machine? We instinctively say no! Yet mathematically, if we have two entities, and for every element of one entity there is one and only one corresponding element of the other entity (and vice versa), we call those entities *isomorphic*, a mathematically fancy way of saying "the same thing."

So what is the difference between our quantum state "on paper" and the quantum state of the nanomachine?[1] One, we say, is an instantiation (which might be on paper or on a hard disk or in a person's memory) of some set of information. The other we say is "real": made of atoms, or elementary particles.

Yet, what is an elementary particle? As Stephen Weinberg, a titan of particle physics, admits:[2] "Suppose that [a] stranger should ask, 'What is an elementary particle?' I would have to admit that no one really knows." What Weinberg means is that an atom acts just like a particle—it has just a few numbers to define it—until you look very closely at it, and see that it is an *arrangement* of protons, neutrons, and electrons. A proton is a particle until you probe it further and discover it is composed of quarks. Are quarks, or electrons, truly fundamental? Probably not; we cannot probe them at fine enough scales to know.

But suppose they are. Suppose there is no sense in which quarks or electrons are "arrangements of" other particles. In that case, we can still ask: "What *is* an electron?" We almost irresistibly want to think it is a little bit of substance. But it isn't. The most sophisticated particle-physics account is that an electron is an excitation of the *electron field*, which pervades space-time. This is very much like a wave or particle of light being an excitation of the electromagnetic field, or an ocean wave being an excitation of a sea of water. But there is a crucial difference. If you ask what the ocean is made of—what its nature is—the answer is that it is *made of water*. If you ask what the nature of the electron field is, the answer is this: An electron field *is an entity able to create and destroy electrons*! This is a rather displeasingly circular response.

So, then, we regroup and ask again: "What is an electron?" Another answer is mathematical: It is an object defined by a particular set of mathematical "symmetries" that *acts* in a certain way defined by particular numbers when measurements of systems are made. It registers a "leptonic number" of 1, a charge equal to and opposite that of a proton, a spin half that of a photon, and a mass 1/1,836 that of the proton. And it has no other attributes.

But in turn, what sorts of things are those attributes? They are answers to questions that are asked, nothing more. There are all manner of possible quantum states, and those of a particular class, with a particular set of possible answers to questions, we call quantum states of electrons. But that, in a real sense, is all there is to it. There is no way to "get at" the electron other than to ask questions whose answers identify it as an electron. Quantum theory strongly suggests that there is nothing *beyond* the quantum state—nothing that the quantum state is *missing*, no *substance* that it is describing. And the quantum state represents potential answers to questions, and really has no further content. In this sense an electron is a particular type of *regularity* that appears among measurements and observations that *we make*. It is more pattern than substance. It is *order*, but of the most clean and crystallized variety.

Thus we arrive at a strange place. We break things down into smaller and smaller pieces, but then the pieces, when examined, are not there. Just the arrangements of them are.

What then, are *things*, like the boat, or its sails, or your fingernails? What *are* they?

If things are forms of forms of forms of forms, and if forms are order, and order is defined by us (who define macrostates) and by history (which actualizes them) and by the Universe (which undergirds the order), then those forms, it would seem, do not exist in and of themselves. They exist, it would appear, only as created by, and *in relation to*, us and the Universe. They are, the Buddha might say, emptiness.

45

THE CLEAR BLUE SKY

(INNER MONGOLIA, 1614)

Intellectually, you know that the sky can't actually be any bigger here than at any number of other places you have been; it can cover only about half of the sphere surrounding you. But sometimes it surely feels more vast, stretching endlessly from horizon to horizon over rolling grassland.

You remark upon this as you take a rare rest stop. Your Mongol guide responds as he, too, gazes up: "It is said that in the clear blue sky, all earthly things exist also, but in perfect, eternal form. We return there, when our time here is done."

This intrigues you. "What are they made of, these clear-sky beings?" you ask.

"They are, well, not made of anything," he replies. "They are the pure essence of the thing, divorced from its material form."

You are reminded of ideas held by the Greeks that you and Galileo discussed at length. You ask: "Do you also hold that there are many things in the great blue sky that are *not* reflected here on Earth?"

"Hmm. There must be," agrees the Mongol, "since I can imagine many, many things that seem to be just as real as those I have seen reflected here on Earth."

"I think my mentor, Galileo, may hold the same. He talks of a 'Book of Nature,' that describes how the world works, and believes that the world changes, but the rules in that book do not." You turn to the monk in your party, who has accompanied you all the way from Ganden. "Do you think Tripa Dragpa would hold that the dharma itself is eternal and unaltering?"

The monk smiles. "He used to hold that to be true, but then he changed his mind."

A while ago I discovered a theorem . . .

—Leonhard Euler[3]

In 1687, Newton first published his universal law of gravitation; 228 years later, Einstein released an update.

In 1748, Euler first published his beautiful identity $e^{i\pi} = -1$.

In the early eighteenth century, Hakuin Ekaku painted *Blind Men Crossing a Bridge*.

In the late nineteenth century, the Zukertort chess opening was developed by its namesake, who curiously did not employ it in the 1886 world chess championship, in which he was defeated.[4]

In 1897, J. J. Thompson demonstrated the existence of a negatively charged particle, the electron.

In 1992, the first extrasolar planets were identified through observations of a pulsar.

In 2005, an unknown aria written by Bach in 1713 was found and verified as authentic.[5]

In which of these cases was something *invented*, and in which cases was something *discovered*?

We tend to think of discoveries as events in which, after the discovery, we feel that the thing had already been there, waiting to be happened upon. Bach's lost aria feels this way, as do the planets around PSR B1257+12 and the electron. Inventions—or creations—newly exist because of the efforts of their creator. Hakuin's painting and Bach's original writing of the aria are prototypical examples. Yet, other quite crucial cases are much less clear. Did Euler's formula exist before he wrote it down? What about Newton's gravitation law, or Einstein's field equations? Or Zukertort's chess opening?

A strong argument for Euler's formula or Newton's law being *discovered* is that these statements—one about complex numbers and one about gravitation—seem clearly to have been true before their discoverers, or indeed *any* people, were around. Could $e^{i\pi}$ have been anything else? Once it is shown to be −1, we immediately conclude that it always has been, and always will be, −1. Likewise with Newton's gravitation law. It may not be a perfect description of the world (Einstein's is better, and perhaps something better yet exists), but we feel confident that planets followed it with extreme accuracy for billions of years before Newton knew about it.

All this suggests a sort of three-way division, made famous by Karl Popper,[6] of *things* in the world. In one category would be actually existing objects and events like chairs, books, and the last game of Zukertort versus Steiglitz.

In a second category would be the world of human-generated informational objects, like all of the mathematical theorems we have proved, works of literature, sonatas and cantatas, and the sequence of moves in Zukertort versus Steiglitz Game 10. In these cases the information could be instantiated in many ways—in books or on paper or via computer memory or by neuronal arrangements—and represents a selection of something *interesting* out of a huge possibility space.

A third category would contain a smaller subset that is "objective" in some sense: laws of physics, true theorems of beautiful mathematics, and other discoverable, abstract things. A sort of Platonic realm—the clear blue sky that exists independent of any physical instantiation.

This categorization, however, draws lines that become more and more blurred, the closer one looks. Consider the phrase "actually existing." What does *that* mean? What does it mean to *exist* in the "earthly" realm (to use the Mongol's term), and to *be something* rather than, say, something else?

The original copy of the painting *Blind Men Crossing a Bridge* now hangs in a museum. We could say that this object owes its "existence" to having solidity, energy, and so on, or perhaps, in more concrete physicist terms, to being made of atoms. The "being something" would then be tantamount to the *arrangement* of those atoms; from this form and the properties of the atoms follow properties of the painting such as its shape, scent, weight, color, and so on.

Why base existence on atoms? Partly because of their *durability*—their propensity to exist over time. And underlying this propensity are what physicists would call *conserved quantities*. Energy is the most well known, and some understanding or intuition of the conservation of energy is what most people would point to as the reason that "things can't just appear or disappear." This is especially true because Einstein's famous equation $E = mc^2$ revealed that energy and mass are interchangeable, so conservation of energy includes conservation of mass as well.

Energy is only part of it, though. Protons and neutrons are in a class of particles, called *baryons*, that are made up of three quarks. Like its energy, the total number of baryons in a physical system is conserved over time. Without this conservation requirement, two neutrons, for example, could simply meet and annihilate into two photons; a proton would be similarly unstable. Ordinary matter would very quickly evaporate into a storm of heat and light. A similar conserved quantity, *lepton number*, prevents electrons likewise from annihilating with protons. Conservation of another quantity, *total electric charge*, prevents electrons from transforming into neutrinos.

We're quite used to how these rules manifest themselves in the everyday world, where objects and substances are enduring, if malleable, in form, and these rules are the bread and butter of particle physicists' understanding of what particles can interact with what others and in what way. But they are not sacrosanct. Energy conservation of the familiar type can be flagrantly violated in general relativity. For example, in an expanding universe the energy of photons like those in the cosmic microwave background simply "goes away" as the universe expands and the radiation cools, and there is no good answer to "where does it go?"[7] In the standard model of particle physics, baryon number is slightly nonconserved at high enough energies (as existed in the very early universe). It is also blatantly violated by black holes, which can form from baryons but evaporate[8] into radiation and particles with zero net baryon number; with a small enough black hole you could destabilize everyday matter at will, converting it into radiation.*

Moreover, we've already seen that the very notion of *particle*, which feels quite solid in classical physics, becomes rather fuzzy in quantum mechanics and especially in quantum field theory, where particles are excitations of fields (which in turn are propensities to create particles, and so it goes around). In fact, it turns out that in *relativistic* quantum field theory, which marries quantum field theory with Einstein's relativity, different observers may flat out disagree about which particles are present.

And finally, we've seen that when we step back to the world of large collections of particles and macroscopic objects, a single quantum state can correspond to *multiple* objects: through quantum evolution and decoherence, an object can split into two or more versions that are in different locations or that differ in their details.[9]

All these things are useful to keep in mind when you hear statements like "it's impossible for something to come from nothing," which tend to be quite misleading unless there is precise agreement

---

* As with nuclear chain reactions, we are fortunate that processes violating baryon number are so hard to get started, since although they could be extremely useful as energy sources, they could be just as extremely dangerous!

on exactly what "something" and "nothing" are supposed to mean. (And there almost never is!)

SO, *to physically exist*, even in a fairly hardheaded physicists' built-of-atoms sort of way, is quite a bit more slippery than one might generally think. What about *to be something*, if we correspondingly think of it as the arrangement of atoms? Even more so!

We've seen already that the arrangement of atoms is "made of" what we've called (structural) information: the particular configuration or set of configurations relative to the full possible set of configurations of the same set of ingredients. This is useful, but we quickly find a rather hazy line between the existence of this structural information and the existence of a particle, insofar as a particle is an arrangement of other ones, or a structure in a quantum field, or equivalent to a quantum state.

Moreover, if we consider this information either quantitatively or qualitatively, we see that it becomes very hard to keep things wholly objective. Quantitatively, recall that we've identified *information* as a lack of "randomness" (which is in turn defined by the vagueness of the probabilities describing the state of the system), and *order* as a lack of "disorder" (which is in turn defined by how generic the state is when blurred out into a higher-level macroscopic description). Both of these quantities, though, depend on probabilities (which are at least partly arguably subjective), and *order* also depends on the definition of macrostates chosen by an observer to use in describing the system.

Beyond how *much* information there is, we may ask about the *qualities* something has that are somehow connected with that structural information. These qualities, which we might regard as making up the essence of *what* a thing is, are fairly clearly *not* objective properties of the system independent of the observer regarding them. Where, for example, does "greenness" reside? No individual atom is green (atoms are smaller than color!). Collections of molecules can have a property of preferentially reflecting particular wavelengths of light. But the word "green," the feeling of seeing green, the associations of green with plants, and so on—all of these exist only in relation to someone seeing a green object with a particular set of color receptors. It's not at all clear that green is a meaningful property even to other animals with eyes (especially if they have two, or four, color receptors rather than three), and it's quite clear that without any biology, "green" would not really be a thing. This is not to say that emeralds are *not* green when people aren't around or that the wavelengths of light emitted by beryllium aluminum cyclosilicate changed when people arose, but rather that greenness itself had to wait.

Similar considerations apply to all manner of properties that we might ascribe to things, including not only those that we are accustomed to having a subjective component, such as utility and beauty, but also, if we really get down to it, those like solidity and sharpness: most or all humans might agree on them, but just like green, without any humans are they really there?

THIS UNAVOIDABLY SUBJECTIVE component of *whether*, as well as *what*, things exist gives the question of discovery (as opposed to invention) a somewhat different tone, since it is awkward to describe something as "discovered" when it is tied up with the discoverer in a manner not wholly different from the way an invention is.

But what about discoveries, such as Euler's theorem, that seem to exist independent of any physical system and seem *to have existed* long before any observers were around to define, talk about, or invent them?

Well, Euler's formula is true (and was and will be true) because it can be *proved*. That is, a set of mathematical axioms and rules has been formulated, and by the application of those rules to those axioms, Euler's formula and many others have been derived. Whether the *axioms* were discovered or invented is rather less clear. Particular choices of axioms appear to bear rich, useful, and interesting mathematical "fruit." But we can imagine that alternative sets of axioms could also lead to a rich and interesting structure; and even if the structures implied by other sets of axioms were boring or impoverished, would that make them any less "discovered" than those we use?

Moreover, the same set of axioms and rules could give rise to a truly enormous amount of mathematics. Imagine the djinn setting up some Djinnium to simply spit out all possible statements derivable from a standard set of mathematical axioms. It would churn out line after line of mathematically correct statements. But would it be *doing mathematics*? It's hard to imagine a mathematician saying so, any more than a writer would say that monkeys banging on a typewriter are authoring.

After all, once the axioms and rules are determined, the set of all statements derived from them is simply a well-defined (if complicated), infinite set no different, in principle, from the set of all sequences of moves on a chessboard, or strings of words. However, *almost all* chess move sequences, strings of words, or mathematical statements are completely uninteresting! There is a minuscule subset that *is* of interest, and finding this tiny subset is, in a real sense, what the creative process does. Indeed, automatic theorem-proving computational systems exist but are not all that useful. A very simple math proof, such as showing $2 + 2 = 4$, starting from $1 + 1 = 2$, employs 10 lines, 26 axioms, and 40 definitions,[10] and a vast, vast number of valid statements would have to be tried out before that result would be happened upon. For such a system to ever find, for example, Andrew Wiles's 108-page proof of Fermat's last theorem, would probably take a metakalpa. Just as composers choose very special sequences of notes, and competing chess masters pick out fascinating, hard-fought games, mathematicians pick out very select chains of reasoning to prove theorems that are interesting and beautiful. In doing so, they generate information in just the sense we saw with the Lhasa River: out of all of the many, many possibilities, only a tiny fraction are pointed to.

From this standpoint, the fact that $e^{i\pi} = -1$ *was true before humans were around* seems rather unimpressive; so were countless boring mathematical statements, all possible chess games, and so on. Once we define a set of rules, we can see that all of the logical possibilities defined by those rules *always have been and will be* defined by them. Why, then, do mathematicians but not composers tend to feel that they are discovering things?

A plausible explanation is that, at least in its roots, mathematics, like physics, was formulated to be in correspondence with the physical world. There are many, many mathematical structures that simply don't have any utility in *doing* things like counting or understanding or predicting, and very few that do. These few can thus be discovered in a way comparable to how the electron was discovered, as part of the nature of the universe. It's not clear that this explanation really captures what mathematics feels like to mathematicians, though; it *feels* as though not just all theorems but the *specialness of certain theorems* is "objective" even when those theorems are quite abstract and feel fairly divorced from physics or the material world. It also feels—and this is a bit hard to convey without actually doing a lot of physics in practice—that there is a mathematical "reality" in the world of physics that *pushes back* as much as a wall does when pushed upon. It is unceasingly amazing to me that as a physicist, one can take a situation in the real world, represent it mathematically, then *do a bunch of pure mathematics*—lines and lines and lines of it, largely forgetting about the physical world entirely—then translate the mathematics back into a statement about the physical world. And that statement will be *true*.

IT THUS SEEMS that whether something is an atom or a theorem, there is some sense of being "made of" information to at least a very large degree, and of the quantities and qualities of that information being at least somewhat determined by *who* has it, defines it, or creates it.

Nonetheless, it feels very much that some objects are special, transcending space and time in some eternal blue-sky sense.

But which? What things belong to the clear blue sky?

Does *the wrongness of killing*?

Do *circles*?

Does *the method of logical reasoning*?

Does *the fact that the djinn's un-run Program 364,343,234 halts*?

Or *the fact that theorem X correctly follows from axioms Y*?

Just how beautiful must a theorem be to reside in the clear blue sky?

46

AT THE FOUNDATION

(HIRADO, JAPAN, 1620)

The Dutch trader ends his unbelievably long, complicated, hard-to-follow, and boring description of the rules governing Nagasaki with a disgusted "only the Japanese could create a system like this."

You've recently arrived and are trying to get the lay of the land. The rules seem like more of a joint Japanese-Dutch creation to you. "Is there something particularly Japanese about complicated rules?" you ask.

You're then treated to a lengthy disquisition about the foundation of the Japanese character lying in the Samurai code, and Shinto, and Buddhism. You finally get a word in edgewise: "But surely Japanese society also affects the character of its people, as much as the people's character affects how society is set up."

He'll have none of it, and continues his exposition with an explanation of the role of the Zen temples in politics. Your mind wanders, and you keep coming back to the *foundation*. What is the foundation of a people? Of a person?

What is at the foundation of the world?

Physicists are fond of thinking of their discipline as "most fundamental"; even more so particle physicists, relativists, and cosmologists. Ernest Rutherford famously said: "All science is either physics or stamp collecting." This statement partly expressed an uncommonly dismissive attitude toward other sciences as being more empirical, less effective and powerful, and less grounded in basic sound principles. But it also expressed a much more commonly held view: that physics *underlies* most other sciences. That is, were we just clever enough, we could start with physics and deduce chemistry, biology, astronomy, sociology, and so on. Even if difficult or impossible in practice, this view holds that these other disciplines are, *in principle*, derivable from "fundamental" physics.

This word is used all the time: fundamental physics, a more fundamental understanding, fundamental particles, and so on. What exactly is meant by it? In particular, what does it mean for one *theory* or set of rules, or description of the world, to be more fundamental than another?

What physicists, at least, generally seem to have in mind is a particular two-part relation. Suppose there are two theories, F (for "fundamental") and D (for "derived"), that both in some sense talk about the same system. Suppose further that once you specify theory F, then theory D is *fixed and determined*: D could not be any different, once you have specified F. Then, suppose further that the number of ingredients, entities, relations, and so on in theory F is inarguably much smaller than in D: F is a *simpler, more elegant* theory. If all this is true, then we call F "more fundamental" than D.

An oft-cited example is the quantum theory of electrons, protons, and neutrons (theory F) and that of the atoms in the periodic table of the elements (theory D). Once you decide on the three types of particles and how they interact, the elements are determined. There is, for example, an element with 2 protons and 2 neutrons (helium), but no element with 56 protons and 2 neutrons; quantum theory predicts that such an element would be completely unstable. Moreover, a handful of properties of three particles, and a few equations governing them, are much simpler than a big complicated table of elements with many properties each.

On the other hand, one could argue all day (say, with a Dutch trader) about whether the "laws" of human nature, or the charac-

ter of a people, or the (literal) laws of society are most fundamental. Clearly, there is a two-way relation between the social, moral,
aesthetic, and other tendencies of a group of people, and the laws
and social institutions that they create. Both profoundly influence
the other. Even human nature is deeply affected by society, not just
during people's lives but on a genetic basis. Our biological makeup
has been deeply affected by our invention of language and its uses,
by the development of agriculture, hunting and migration, cities,
technology, and so on.

It would seem, then, that some levels of description really are
more fundamental, like descriptions in terms of atoms, but at some
point (perhaps in biology) the hierarchy becomes less clear.

YET EVEN the seemingly clear cases are not so clear when inves-
tigated further. For example, fixing the properties of atoms determines the properties of the periodic table. But could we really alter
the properties of chemistry *without* also changing the properties of
atoms? No! Given an ironclad demonstration that a chemical reac-
tion was *different* from what current theory indicates, we would not
and could not shrug our shoulders and say that "quantum theory
and particle properties (F) determine chemistry (D) but chemistry
does not determine quantum theory." For if F determines what D
is, then it also determines what D is *not*! So we would be forced to
make modifications to F to create a new theory that would accom-
modate our new understanding of D. Indeed, this is just what we do
when we test "fundamental" theories. But if this is so, then we have
to admit that while theory F has the virtue of greater simplicity, it
has no greater *determinism* than theory D has.

Moreover, the apparent determinism often glosses over ingredients
that are quite vital. Consider the *laws of heredity* that apply biologically and are specified in terms of genes, chromosomes, reproduc-
tion, and so on. Chemistry is clearly of central importance. But the

genetic laws could almost certainly not be *derived* from chemical (or atomic) rules. The use of DNA, the coding scheme with amino acids (which are left- rather than right-handed), pairing of chromosomes and sexes, and so on, are, in very large part, historically determined. They work well, but there are almost certainly other "solutions" that might underlie genetics and heredity that are quite different. Thus, to go from a putatively "fundamental" chemical theory to a "less fundamental" genetics description requires a large amount of auxiliary historical information. Where does that information fit into the scheme? If you add it to the "fundamental" chemical description, then that description does not seem so simple anymore. In addition, given the rules of genetics that we have, it might be quite hard to find a *different* set of chemical or atomic laws that could give rise to those rules, no matter what the history of evolution. So, the direction of determinism probably points *more* the other way, from putatively "less fundamental" toward putatively "more fundamental"!

Surely, though, we can avoid such pesky issues and salvage the pride of the particle physicists if we stick to physics and chemistry? That's still unclear. Recall from our contemplation of THE PAINT-ING IN THE CAVE that what we think fixed, like the laws of atomic physics, "may change, dreamlike" among different universes. This is because "one dharma" that "pervades them all" manifests as different rules that apply at lower energy scales. Thus, there could be a one-to-many relation between one encompassing theory and the many possible sets of "fundamental laws" of the type that particle physicists study in their collider experiments. In this scenario the laws we study would be contingent on our particular cosmic history, or perhaps *where we are* in some cosmologically generated multiverse that holds all possibilities.

Other aspects of basic physics might also be more contingent than we think. Were we ever to encounter and understand extrater-restrial life, or develop artificial intelligences powerful enough to formulate laws of physics from a wealth of data, would they develop the same physics? We can guess that their physics would be as effective, and it seems likely that there would be some level of mathematical equivalence to ours. Yet it could be conceptually *very* different, based on quite distinct ingredients. Textbook treatments of quantum mechanics, for example, are generally based on Schrödinger's *wave mechanics* and Feynman's *path-integral* formulation (both of which we've studied here). But there is Heisenberg's equivalent *matrix mechanics*, Bohm's *hidden variables* theory, and formulations (at various degrees of success and completeness) in terms of category theory, information theory, constructor theory, cellular automata, and more. Even if mathematical mappings between formulations exist, those mappings are often not complete, and different formulations might lead to quite different worldviews and open up different avenues that invite further exploration. What made Schrödinger's mechanics dominant, at least in textbooks and college courses? Probably no small part was simply that it was among the first, and it made calculations easy.

Thus, in many physical systems that we would like to describe, there is a combination of both a set of relatively simple rules, and also some rather complicated specifications stemming from initial conditions, historical contingency, specification of where we are in the universe (or Universe or multiverse or quantum state), the social history of what concepts are most natural for us to apply, and so on. The rules governing the system are an irreducible combination of both ingredients. And when both are included, the idea of *deducing* what we think of as less fundamental descriptions from more fundamental ones becomes rather trickier, with the direction of implication potentially running in the opposite sense.

DESPITE ALL THIS, it still seems that something is more fundamental about atoms that goes beyond the fact that they are small:

the *conceptual simplicity* they enjoy, as opposed to the messy, complicated world of biochemical pathways or ecosystems or Bushido.

Here too, though, there are countervailing examples. Sometimes the rules governing systems made up of many, many particles can be beautiful and simple. Consider the relation of *computation* to atoms. Computation has an elegant, well-worked-out theory in terms of Turing machines, logic gates, and other entities. These ideas can be applied whether the elements of computation are transistors, Legos, black-hole states, DNA base pairs, or Djinnium. The same is true of the laws of statistical mechanics, such as the second law of thermodynamics: one might as easily apply them to monastery kitchenware as to atoms.

These laws are incredibly strict. If you claim you have a device that violates certain laws of computation, let alone the second law, nobody will or should take you very seriously. The formal proofs and theorems governing computation feel rather objective in the same way that mathematical physics does. And yet there is something unavoidably human-dependent (or at least observer-dependent) in them. Without human-created systems like transistors and microchips, the rules governing logic gates, Turing machines, and computational complexity would be quite irrelevant. Similarly, the second law, of disorder increase, feels like a powerful and foundational rule of nature, yet *we* define order through our definition of the macrostates, as well as our definition of the "system" that is being discussed in the first place.

So where are we? It very much *feels* that there are "fundamental rules" to the physical world; they are the things that physicists study. There is truth to this. But it is far more subtle than it initially appears. We physicists sometimes say that such-and-such is "nothing but" a collection of this-and-that governed by these-and-those. You should take this seriously. Like the Dutch trader, physicists have learned and understood a great deal about how the world works by studying "fundamentals." But also like the Dutch trader, don't take them *too* seriously.

47

THE GREAT INHERITANCE

(HERAT, PERSIA, 1611)

Before leaving Europe you never would have believed how much desert there could possibly be in the world. By now it feels like you've personally walked it all. You have been traveling toward Kashmir largely because you've been promised an end to the desert there, replaced by great beauty.

As you approach a quite lovely oasis of a town, a companion launches into one of his many diverting, if endless, stories.

"This town is often called the Pearl of Khorasan. But few know that there is an *actual* pearl of Khorasan, an object of great importance that was passed down secretly over the ages from ruler to ruler."

"What was it?" you ask gamely.

"Nobody knows. That is, nobody knew what was inside. It was a crystal enclosure of surpassing strength. Strange but obviously valuable. And it is said that anyone, upon touching it, was immediately endowed with utter certainty of a particular truth."

He waits, expectantly.

"What truth?" you say as if dutifully, but secretly your curiosity is piqued.

"The truth," he says, "that the box contains either inestimable value and preciousness, something of utter rarity and worth, *or* that it contains something of ultimate danger, and chaos. It is said that it comes from either the immemorial past or the inestimable future. Apparently, it came into the possession of many rulers over the centuries, but none dared open it."

**You wonder if you would open it. But only for a moment.
Of course you would.**

To see a World in a Grain of Sand
And a Heaven in a Wild Flower,
Hold Infinity in the palm of your hand
And Eternity in an hour.

—William Blake, *Auguries of Innocence*

When you are given a gift wrapped as a present and you say, "I don't know what is in the box," you are not really being honest. From a cosmic perspective, you know, or at least predict with very, very, very high confidence, almost all the information about what is inside the box, save for a few minor details.

In particular, you "know" in this confident, predictive sense that the inside is materially composed of atoms made of protons, neutrons, electrons, and photons, and that the typical energy per atom is less than that corresponding to about 100°C. If all you *really* knew were externally measurable things—say, the mass, volume, electric charge, and so on—about the present, then you would not really know any of this. For all you'd know, the interior could be an ultra-hot plasma. Or a tiny, shining black hole.

In fact, it *should* be something like that! There's a well-known technique in statistics that applies when you have a system with a known set of possible configurations or states but you know only a small number of facts—call them X, Y, and Z—about it. The technique then enjoins you to assume the most *generic*, or *most random*, configuration you can that is compatible with X, Y, and Z.[11] The logic is fairly unimpeachable: since randomness is a lack of information (as we learned from the Lhasa River), maximum randomness is *minimum information*. That's really what "know only" means regarding facts X, Y, and Z. If you had some reason to assume something *other* than the most generic and random state, then that reason would by definition be *information*, which would correspond to a thing W that you know. You should then be clear about that, specify W, and infer the most generic configuration consistent with X, Y, Z, and W.

Applying this method to the wrapped present, if all you *really knew* at some instant of time was about the exterior wrapping, then properly you should infer a maximum entropy state inside. What does such a maximum entropy state look like? Physics provides an answer, and it's the same one we saw when we imagined the thangka painting sitting in its box forever: a maximally random and disordered state would be something like an ultra-hot gas of fundamental particles, perhaps even an evaporating black hole, or the like.

One might object that the wrapping could never contain such an interior, which would instantly explode into an incredibly powerful bomb.[12] This is true. But specifying that the wrapping stays intact for some number of seconds is just a piece of data W that must be factored in. If, say, you know that the box has been stable for 1 minute, then you will infer that the interior is the most random configuration consistent with being at room temperature on the outside for 1 minute. This is probably a very, very hot medium confined by an incredibly effective and strong insulating shell that can last just about a minute. So step back!

This is a road to madness. Of course, we never do this. We never *really* think that the contents of a box are the most generic possible contents subject to our knowledge. You might guess what's inside a gift box by guessing what a "typical" or "generic" gift might be—a Lego set if you're a kid, a necklace if it's your anniversary, a tie if it's Father's Day—but unless you have very unusual friends, you would never guess an ultra-hot plasma! Thinking that way, you would be terribly wrong every time, and disappointed or relieved every time you opened a box. But *what* exactly is wrong with thinking that way?

Maximum entropy estimation is neither silly nor wrongheaded; it is almost by definition the right way to characterize the state of something about which you have limited information.

Rather, what has gone wrong is that in the everyday world, you have a *huge* amount of *implicit* knowledge of objects, like the gift's interior, even while professing ignorance.

This knowledge is bequeathed to us by the extremely highly ordered, information-rich state of the early universe. The information-rich universe created highly ordered galaxies, inside which information-rich stars formed with high-information planets and environments around them. The information and order are stored in the coolness of matter, the binding of atoms and molecules, the emptiness of space, and the resistance of objects to gravitational collapse. It doesn't *feel* like an amazing store of information, any more than a refrigerated burrito does. But it is a secret, crucial, and (from the standpoint of the universe stretched throughout space and time) incredibly rare resource.

Everything you see around you has a billions-of-years shared history that allows you to *infer*, from a tiny amount of data—some photons in the eye, some vibrations in the ear, some nerve signals in your finger—an absolutely astonishing amount of information. It feels as natural as breathing, because we're so used to it and think in terms of the information we have relative to the information we'd have if we (say) peeked inside a box. But this extra bit of information—a bracelet, not a necklace—sits atop an astounding amount that we already knew, because we were born into it. Over millions of years we have evolved to make use of this implicit knowledge so as to understand, predict, and manipulate the world in a way that would never be possible if it were truly chaotic or random.

This vast knowledge is not all that we have inherited from the universe. The growth of disorder provides an *arrow of time* that points from the highly ordered early state of our universe toward what is then called the "future." This arrow points in the direction in which the description of the universe branches, and branches, and branches into many possibilities. It underlies our ability to affect the future very, very much more than the past and to record the past very, very much more reliably than the future. It is thus, in a real sense, *time itself,* creating the current that underlies our existence and experience as beings.

THE RICH STORE of information combines with the creative historical process of generating order through time to form what is perhaps the most valuable inheritance: the incredible complex structure of living things on the *physical, mental,* and *emotional* levels.

On the physical level, biological creatures are so much more complex *in a functional way* than current artifacts of our technology that there's almost no comparison. The most elaborate and sophisticated human-designed machines, while quite impressive, are utter child's play compared with the workings of a cell: a cell contains on the order of 100 trillion atoms, and probably billions of quite complex molecules working with amazing precision. The most complex engineered machines—modern jet aircraft, for example—have several million parts. Thus, perhaps *all the jetliners in the world* (without people in them, of course) could compete in functional complexity with a lowly bacterium.[13]

On a mental level, the systems that bring glints of photons into awareness are just one of many that understand sensory input as part of a hierarchical mental structure with multiple levels of feedback that are constantly perceiving, modeling, predicting, comparing, sorting, optimizing, and acting. The intimate familiarity of these systems should not belie their wonder. Researchers in artificial intelligence have long been acutely aware of how monstrously difficult it is for computers to compete with biological minds on their own territory—not of numbers and bits, but of perception, prediction, and action. Perhaps even harder is to operate comparably on the level of contextual understanding: to conform perceptions, predictions, and actions to the vast standing repository of concepts, understandings, prohibitions, and constraints that make up the biological and social world. This repository, as well as our intimate connection to it and ability to function within it, is a precious inheritance and a store of much of what we value.

But at the *emotional* level, the level of awareness and even spirit, we've been bequeathed probably what is most precious of all, and what really makes us each AN HONORED GUEST. After all, what does anything *matter* if nobody is aware of it, conscious of it, feeling one way or another about it? We value things almost entirely because of their effect on the feelings of people and other living beings. Most wrong actions are wrong because they cause *suffering* in one form or another, and good deeds are generally considered good ultimately because they aid in people's feeling of well-being. But the range of feelings goes far beyond their being *positive* and *negative*: they make up the core of what is meaningful about being human. If we had distinct words, rather than the paltry handful like "happy" and "painful," to describe them, it would take thousands. Instead (though probably to everyone's benefit), we resort to poetry and complex prose just to try to convey or point toward the feelings we hope to share with others.

Some dismiss feelings like romantic love, or awe, or nostalgia, or even pain or anger, as "just brain chemistry" that has evolved as part of the Darwinian process. This is, of course, untrue: inside a bottle, serotonin and dopamine don't have feelings; they are *signals* in the brain that coordinate with existing structures and processes corresponding to these complex feelings. Moreover, imagining that these things evolved through an evolutionary process in no way *cheapens* these experiences or makes them mean less. To the contrary: what could be richer than something that has been built up over millions and millions of years and generations over the entire Earth and a vast number of organisms? When we contemplate designing machines to have such feelings, it's baffling even to think about where to begin.

These feelings are intimately intertwined with our bodies and intellect, and all three have developed through what we might think of as a world-spanning computation, or a vast mental process stretched over time and space in AN UNFETTERED MIND.

It is hard to imagine anything richer, or deeper, or more complex than this.

48

A LONG HIDDEN GAME

(EDO, JAPAN, 1629)

Munenori and Sōhō are sitting at a game of Go when you enter. They ignore you.

As you watch them, you grow increasingly uneasy as you realize they are breaking the rules. Or at least not playing Go. During a momentary lapse in the room's focus, you ask: "What is this game you play?"

"That is for you to understand," replies Sōhō. You almost think you see Munenori smile, which always signals danger.

The game wears on, each move taking longer than the last. Try as you might, just when you think you understand the rules, a played stone proves you wrong.

At last, Sōhō concedes: "I have no move." Munenori nods a bow.

After a time, Sōhō speaks: "Sometime I would like to hear the rules you played, in the end."

You are surprised. "You didn't know the rules?"

Munenori turns to you. "Neither knew the rules in advance. So is the world."

"And yet you won."

Munenori resolves: "So it must be, with that beast in the cave, and its infernal dove."

You nod, and the three of you sit together, resuming where you left off the day before, discussing the plan.

> When we were out of the room, they explain, they
> had agreed not to agree in advance on any word
> at all.
>
> —John Wheeler, recounting a fictitious parlor game[14]

When we ask "what is in the box?" we are well accustomed to assuming that there is an answer to that question not just before we open the box, but before we even ask the question. A box might contain a chunk of coal or a diamond, but no amount of wishful thinking will affect what is discovered upon opening the box.

Yet this assumption that the answer already exists apparently fails, as we now well know, when we go smaller. If someone places a solitary neutron in a box and hands it to you, the question "will I find a neutron when I open the box?" simply has no answer until you actually open the box. This question might have a definite answer ("yes") at the moment the neutron is inserted, but the quantum state of the box's interior steadily evolves into one containing both the possibility of a neutron being there and also the possibility of the electron, proton, and neutrino (into which a neutron naturally decays) all being there. When you open the box and ask your question, you force nature to take a stance, and nature does so. Through this process you affect the system in a way that is only partly under your control: you choose the question but cannot control the answer you receive. That is, you choose a set of states that the system can be placed in, but the system chooses *which* state.

This "participation" in creating the content of the box might seem a curiosity to be relegated to the quantum world, except isn't everything made of quantum stuff? Yes, it is. And given quantum theory, the way to properly talk about *macroscopic* systems is again as a set of questions to be asked of those systems. We choose the questions, and nature provides the answers. There is no clear, bright line, though, demarcating when the answer to the question exists before we ask it (as it seems to for coal and diamonds) and when it does not.

So, then, is the world made up of answers to questions? And if so, how much of it is construed from the questions we choose to ask?

In a famous paper on the implications of quantum mechanics, John Wheeler recounted a parlor game in which a group of guests agrees upon a secret word; then a questioner, not privy to the discussion, enters and asks questions of several guests in turn, so as to zero in on the secret. But in Wheeler's version, the group only agrees *not* to choose any word. Rather, as the questioner asks more questions, each respondent must select a word that is consistent with all of the previous answers to previous questions, and base the answer to the next question on this chosen word. It is quite clear in this game that the properties of the word are a co-creation of the question asker and answerer, and that there is no meaningful sense in which the particular word or words that might be described exist at the beginning of the game. Wheeler felt that the participatory manner in which the words came into being captured a deep truth about reality, which he summarized in two aphorisms: First, "it from bit": existence *is* information—in particular, information that quantum systems provide by answering questions put to them. Second, "law without law": there is no single preexisting set of physical laws; rather, they come into focus more and more as we see and understand more of the Universe through our question asking and answering.

How seriously should we take these ideas? It's tempting to dismiss them if we think only of *people* asking the questions, as in the parlor game; surely, stars, galaxies, and many other things existed in some sense before there were people around to ask questions about them, and the Universe did not somehow pop into existence when the first person asked about it.

People, though, are not the only entities capable of asking questions. Any physical system can "ask a question" in the sense of interacting with a quantum system (which is to say, any other system) in such a way that the interaction leads to multiple distinct, decohered outcomes. We can thus imagine a sort of hierarchical structure in which "things" ask questions of smaller/more quantum "things" and, in doing so, create information and also *shape* the world via the *choice* of questions asked. This choice, of course, is only a choice *as such* for systems that can be said to choose, such as biological systems that came relatively late to the cosmic scene. But anytime a relatively complex system interacts with another, so that there are many possible outcomes at the *system* level of description, this is the asking and answering of a question. In this sense, atoms can ask questions of electrons and protons, molecules of atoms, dust motes of molecules, planets of dust motes, RNA strands of particular organic molecules, proto-cells of RNA strands, and so on. Through this process, information and order are generated.

And while the growth of order we saw over the Lhasa River allows the intricate structures making up our world to come into being, it does not specify what those structures *are*. We saw THROUGH THE LOOKING GLASS that some of them, like stars, are well determined by the basic physical equations and constants. Other complex systems, like living ones, might very well be described as intimate, never-ending, multilevel back-and-forth games of 10^{20} questions in the cosmic parlor. Moreover, as we saw as dishonored prisoners in Tripoli and honored guests in Agra, adding the crucial element that we, in talking about any of this, are conscious, complex, information-processing observers places a profound *postselective* effect on the long list of questions. We minds only get to ask questions that are at the end of a very, very long series of more implicit ones that include affirmatively answered questions like "am I sentient?" and "am I a thinking being?" If we bring the world into being by asking questions, and if the chains of questions have to end in questions like these, isn't it a very, very, very special type of world that we have brought into being?

But are we really so powerful? Under THE CLEAR BLUE SKY we may well ask: "Aren't there things that are not made historically, but rather *baked into* the fabric of the Universe?" Don't we *discover* mathematical theorems and laws of fundamental physics that have been there all along, waiting for sufficiently wise and clever beings to uncover them? Even if it takes thinking beings to conceptualize and write the laws down in language and mathematics, and even if those mathematical structures are created by our interest in them out of all possible sets of structures generated from all possible sets of axioms, nonetheless did those laws not *exist* in some way to tell the universe how to evolve? How could the rules themselves come into existence? What rules would govern *that* process?

The existence of *some* set of fixed, discoverable, and immutable laws then seems somewhat inescapable. And yet . . . what would the world look like if there were *not* transcendental mathematical rules governing it?

There is a mathematically rigorous and quite beautiful theory, developed by Ray Solomonoff[15] in the 1960s, called *universal induction.* In very brief form, it describes a completely general way to predict the next item A_{N+1} in a sequence of items $A_1, A_2, \ldots A_N,$ supposing that the list was generated by some unknown process. Universal induction then says that you should consider *all possible algorithms* that lead to the sequence $A_1, A_2, \ldots A_N,$ and aggregate what they predict for A_{N+1}. We convert these predictions into a probability for each possible value of A_{N+1} by seeing how many of the algorithms predict that value, but also *weighting by the simplicity*[16] *of the algorithm.*

You are then mathematically guaranteed that even if you don't know, and don't ever figure out, the unknown process underlying the sequence A_1, A_2, . . . , nonetheless as the sequence gets long, your predicted probabilities will become more and more accurate. Thus, Solomonoff's universal induction does a great, and very general, job at "understanding" the world and exploiting its regularities to make predictions. This sounds like a bit of a miracle: how can you make incredibly accurate predictions without knowing what the *actual* rules are? But consider a black box that spews out bits, either 0 or 1, according to some algorithm unknown to you. One thing you can do is count up the total number of previous 0s, and the number of previous 1s, and predict that the probability of the next bit being a 1 is the same as the fraction of *previous* bits that were a 1. If $N_0 + N_1$ is large, you will have a predictive handle on the next bit, though you'll get the answer wrong quite often. Universal induction, a more powerful version of the same basic idea, can produce much more detailed and accurate predictions.

To be clear, this is *not* how physicists or anyone else actually does things. We don't map all of our previous observations into strings of bits, and it's obviously impossible to run through "all possible programs." So whatever it is that we humans are doing when we try to abstract out mathematical laws governing the universe, it is not really Solomonoff induction. But it may resemble it, or be some evolution-crafted, very clever set of approximations to it. In this case, we might well expect that no matter how complex the world is, as long as it is not utter chaos, then the rules that we generate to explain it will be rather simple, especially if we allow ourselves to cut away much of the complexity (as physicists do when discussing fundamental particles). And would not the mathematics we develop be particularly and elegantly applicable to the relatively simple "explanation" that we devise for the relatively simple systems left after this complexity chopping?

To think of a world that has some sort of *order* to it but, as Wheeler suggested, "no laws"—just as there was "no word" in the minds of the guests at the party—is quite hard.[17]

It's unsettling, perhaps, to imagine that there is nothing at the bottom.

But also, perhaps, freeing.

49

THE MIND-ONLY SCHOOL

(NEAR SHANGHAI, CHINA, 1619)

Since your illness, you have had occasional spells in which you have difficulty telling what is real and what is not. You've also heard stories about divinities appearing on Laoshan mountain. Whatever its reality, you are having trouble understanding what you are seeing and hearing here in the dark, yet somehow comforting, mists.

A strangely compelling Indian man in monk's robes instructs what appears to be a well-dressed British clergyman in an unfamiliar tongue that you—and the clergyman—can somehow understand: "When the Tathagata speaks of 'atoms of dust,' it does not mean that he has in mind any definite or arbitrary conception, he merely uses the words as a figure of speech. It is just the same with the words, 'the great universes,' they do not assert any definite or arbitrary idea, he merely uses the words as words."

As you ponder this, the Englishman queries the monk: "Doth the *reality* of sensible things consist in being perceived? or, is it something distinct from their being perceived, and that bears no relation to the mind?"

He is, in return, answered by a second Indian monk: "Everything is real and is not real, Both real and not real, Neither real nor not real. This is Lord Buddha's teaching."

You're not sure what to make of this, but it appears to trouble another European gentleman, who objects, in somewhat scornful French: "I will suppose that the sky, the air, the earth, colors, figures, sounds, and all external things, are nothing better than the illusions of dreams."

A German of melancholy aspect interrupts the Frenchman, whom he calls René: "It must be so in order to be merely honest. For nothing is more certain than that no man ever came out of himself in order to identify himself directly with things which are different from him."

While compelling, this argument feels incomplete to you, but your thoughts are interrupted by a second Englishman, dressed in strange clothing, who explains: "The mind-stuff of the world is, of course, something more general than our individual conscious minds. . . . Only here and there does it rise to the level of consciousness, but from such islands proceeds all knowledge."[18]

At this, as a group they turn to you, expectantly. "What is not in *your* mind?" the clergyman asks.

You wonder about what is happening. Is it all in your mind? Is *it all* in your mind?

When the Buddha said, 2,500 years ago, that conceptions of "dust" and even "universe" are illusions, what did he mean?

It is difficult to know exactly: the words in question were probably spoken originally in Magadhi, passed down orally for decades, written down in Sanskrit, then translated into English. But they conform to an ancient Eastern (and more recent Western) tradition asserting that the world we experience is something like a dream: largely a construct of our own perceiving mind, rather than something "out there" that exists independently and that we may come along and perceive. This notion predated the Buddha, was carried through Buddhist lineages and especially by the so-called *mind-only* school of thought, probably most comprehensively and provocatively by Nāgārjuna. It has been paralleled perhaps most famously in the West by Descartes's famous meditations on how we can know anything outside the mind, and George Berkeley's famous philosophical response that we cannot, nor do we need to.

In this book we've also again and again seen from the *physics* perspective how what we immediately and habitually take to be real, solid, independent, external objects are, upon closer inspection, much more ambiguous. Objects are made of structural information applied to atoms, which are "made" of a quantum state, which is itself a sort of informational entity regarding what can be measured. And information is "made of" constituents like probabilities and order that exist at least in part in the mind of an observer or perceiver. There is no clear line between entities that are "building blocks" and other entities that are "merely arrangements" of those blocks. Almost all of the things, processes, and patterns that we experience are constructed by physical and evolutionary processes acting on a cosmic endowment of order to craft patterns that we have evolved and labored to understand. Familiar properties like color, utility, and beauty are human constructs that no single atom has, and even more primitive notions like motion, solidity, locality, or change are ways that observers or agents choose to break up and describe the cohesive whole that is reality. Modern understanding of neurology and cognitive science confirms that perception is by no means a passive endeavor, but is a continuous bidirectional process in which the reality we experience is, like a dream, constructed out of the raw materials of sensory input, memory, feeling, awareness, prediction, and pattern matching, moment by moment.

The form of any given thing is really made up of an intricate set of relations and constructs that have coevolved with us and many other things and forms.

Form is emptiness.

But emptiness is also form.

Wishing a door open does not open it, and kicking a stone will hurt your foot regardless of your possible belief in that stone's illusory nature. Will yourself to fly in a dream, and you just may. Awake, you will fall off the cliff: matter can bend space-time, but your mind through sheer will alone cannot. Even clear constructs that we all agree have been invented—money, for instance—are quite real, with their own inviolable rules and patterns. If you try not to believe in money, you can do so for a while, but the consequences will be nearly as dire as failing to believe in gravity. There are many, many such concepts that exist both in an individual's mind and in the collective minds of humanity or beyond. Languages, social conventions, philosophies and religions, political systems, and so on have pretty clearly been constructed by humans though social history, and are learned and reflected in individual people's educated minds. More universal propensities, such as fear of snakes and spiders, sexual attraction, urges both to explore and to remain in safety, and so on are clearly, in large part, evolutionary heritage shared by and across species. The Buddhist mind-only school named these *alaya*, the "storehouse consciousness"; it includes (but perhaps goes beyond) what Jung called the "collective unconscious." There is no denying the causal power or importance (or arguably *existence*) of these objects, even if none can be pointed to in space-time. While rocks, money, and spider scariness might not have an *inherent* existence independent of mind, they're not a transitory pattern of clouds, or a brief vision in a dream either; they are much more durable.

By these lights there is certainly a *somewhat* objective world: things that people agree on in common, sometimes to a very large and confident agree, and that are to be doubted at one's great peril. But the *completely* objective world—that which can be said to exist independently of us or of thought or of awareness or of consciousness—probably contains far, far less than we often give it credit for.

BUT WE'VE ALSO SEEN that the *subjective* is something of an illusion. The sense of individual "I"-ness is also something that we continually and effortlessly manufacture. What you think of as *your ideas* and *your thoughts* are part and parcel of a much larger mental process shared across many beings and many millennia. Even the sense that you are *individually* conscious because you associate yourself with a particular *individual* brain seems problematic. No one has ever satisfactorily clarified or specified which physical systems do, and which systems do not, have an "internal" sense, an awareness, or consciousness. It seems hard to countenance that this is something that is completely binary and either *had* or *not had*; even our own experience of consciousness admits of degrees. Where exactly in the continuum of human, chimpanzee, lemur, rat, lizard, beetle, ant, mite, nematode, amoeba, bacterium, virus, and DNA strand does this quality disappear? What about a colony of ants, the global internet, the community of trees in a forest, or the state of California? Perhaps, as Eddington enjoins, we should "accept the view that the substratum of everything is of mental character" and that "we must postulate something indefinite but yet continuous with our mental nature."[19] This is not to say that an electron, or a rock, has experience or feelings or thoughts anything like ours, or even the simplest lifeform's. Rather it is raising the possibility that not just some systems, but *all* systems, have both a *physical* and a *mental* side to them; that the ideas of *pure* physicality and *pure* mentality both dissolve on very close inspection.

What we call *objective* and *subjective* should perhaps not be considered two sides of a coin, but rather two ends of a continuum.

At one end, your very own things: here and now; your present sensations; your patterns of thought; the way you remember that afternoon 5 years ago; your favorite type of cake; how you love those you love; the indescribable feeling that one piece of music gives you; your regrets over hasty words; your pride at a recent achievement; your despair at what things have come to; your vision for the future of the world.

At the other end, the cosmic and the very smallest pieces of the cosmos: the fact that the ratio of the circumference to the diameter of a circle is given by the infinite sum $4(1 - \frac{1}{3} + \frac{1}{5} - \frac{1}{7} + \frac{1}{9} - \dots)$; Einstein's and Schrödinger's equations; quantum fields and space-time metrics; Turing's noncomputability proofs; an infinite fractal structure composed of bubble universes without end.

But so much in between! Time's unfolding into a succession of instants; three directions of space and one of time; the bylaws of physics; the history of our particular cosmic locality; this branch of the Milky Way's wavefunction; eight rocky bodies revolving with clockwork precision about a fireball; up and down; the structure of life; the laws of genetics; four limbs with five digits each, and two eyes; hunger, thirst, fear, and anger; tools for doing; utterances with meaning; commandments and codes; money; religions; states and laws; works of art; an elegant game of Go; the notes of that one piece of music; hasty words; despair; visions for what the world could be.

And are the ends really at the ends? Could there be some ineffable truth beyond π and proofs and metrics? Could there be some ineffable reality interior to "you" to be glimpsed as time floats away like a spring breeze?

And is it just a spectral continuum, a simple line? What, after all, in all that we have seen, is simple?

50

EAST AND WEST

(DAITOKU-JI TEMPLE, KYOTO, JAPAN, 1630)

You sit across from Munenori over tea in the temple courtyard.

"You think," his words break the silence, "that you have chosen wisely."

You decide this is a question, and as the tea cools and the bamboo lightly knocks in the breeze, you consider it.

You have made so many choices. To leave your mentor and safety for an uncertain and perilous journey. To go east, into a well-laid snare of coincidence. To risk escape into the starry night. To take one more step, again and again in the desert. To rub that infernal lamp. To question, and argue, and argue, and question about free will. To step forward before an empress. To take a cup of chai. To roll the dice. To cross the mountains. To carefully watch rivers and lakes. To aid an emperor. To walk in the mists. To cross the small ocean. To convey a timely warning. To take up the sword and the meditation cushion both. To realize the unfolding game and make one move, then the next. What will you choose tomorrow?

You reflect also on what you may or may not have chosen. Did you choose to survive the road to Shenyang, and come back to the here and now? Did you choose to wake from your dream into this? Did you *really* choose to roll, and cross, and watch, and warn—or were these choices made for you by the Pen, or the djinn, or the dove?

How much of your fate have you made? What will be chosen for you tomorrow?

[People] do not sufficiently realize that their future is in their own hands. Theirs is the task of determining first of all whether they want to go on living or not. Theirs the responsibility, then, for deciding if they want merely to live, or intend to make just the extra effort required for fulfilling, even on their refractory planet, the essential function of the universe, which is a machine for the making of gods.

—Henri Bergson, *The Two Sources of Morality and Religion*

We do so like to divide things up into opposing camps, don't we? Us and Them. Self and Other. Male and Female. Liberal and Conservative. God and No God. East and West.

I can't claim to be any different. Through this book I've presented, generated, practically *assailed* you with, a whole host of dichotomies.

Several involve space and time, and their similarities and differences. We've seen that we can describe a physical system in terms of a succession of *states* from one time to another *or* as a set of possible complete *trajectories* though time. We've seen that space and time are both fundamentally the same interconvertible reality, and also that they are completely different. And we've seen arguments that the present time is *everything* (with the past and future just memories and predictions) and that it is also *nothing at all* (just one arbitrarily chosen subset of space-time).

Closely related to these are conflicts between the *freedom* we experience when we make choices and the apparent *necessity* of those choices that is implied by the inexorable grind of mechanistic laws of physics. Along the same lines, is the true explanation for what happens in the world simply these laws, or is it meaningful to include things like *goals* and *wants* that particular physical systems can push the world toward? These, in turn, hinge in part on whether those laws are *deterministic* (one present leads to one and only one future) or *indeterministic* (one present leads to many possible futures).

Many of the thorniest (and most savory) impasses have arisen between the polarities of the subjective and the objective. Take probabilities, which, if we are honest, are all we really can ever know, given our inherently uncertain world. Are probabilities *subjective probabilities* that pertain to our individual or collective state of knowledge? Or are they *objective probabilities* pertaining to frequencies in an ensemble, or perhaps inherent propensities for outcomes to occur? Likewise, just as everything we know is probabilistic, everything we know about is quantum. Is the quantum state a fundamental reality, which evolves as an ever-bifurcating complex of observers seeing different outcomes? Or is the state *about* some slippery and elusive reality that comes into being as questions are asked and answered? This could be reframed as a question about our journey: do we really take *all possible paths* through space-time, or is that a construct that allows us to understand the *one true path* that we actually do take? These questions take on a poignancy and even urgency when we apply them to our individual subjective existence in a potentially vast and deep and eternal Universe, in which the comfortable, familiar one-to-one connection between personal identity and bodily existence and uniqueness may be radically undermined.

In *information*, the subjective-objective divide found a fruitful ground in which to play out. Is a theorem or an aria or a thangka painting a subjective or objective *thing*? Is information *about* objects, or are objects made up of information? Is an object *nothing but* atoms and their arrangement? Or are atoms *nothing but* particularly simplified isolated and extracted bits of nature? Reality tenaciously and appropriately resists easy answers here: To say that there is nothing objective about a refrigerator seems absurd. Yet we have also seen that upon analysis, everything that we can say about the makeup of the refrigerator has been constructed out of ideas, information, probabilities, choices, point of view, and other such ingredients that lack tangible reality.

Even on the largest scales, we see irresolvable antimonies. Is the Universe finite or infinite, inexhaustible or limited, eternal or transitory, everything or part of something greater? These are age-old questions, not quite answered but surprisingly less pure speculation than they might first seem. The dark night sky speaks to us of the infinity of space and time. Elegant mathematical theories make concrete the largest questions of space and time. The hospitality of cosmic law hints at things beyond the cosmos. The djinn and the quantum gun threaten to reveal mysteries of the infinite, the quantum, and the mind, but at the ultimate price.

To me, one clarion message comes through: when you are tempted to think or say "the Universe is fundamentally like *this*," then go and sit and think some more.

SO WHY do we like to divide things into opposing camps? Probably no small part of it is a rather strong drive, bestowed upon us by our evolution as surviving beings, to identify some certain things with ourselves, our group, our tribe. The "selfing" process walls us off as individuals in an often hostile world and seeks to protect not just our bodies, but by extension, whatever we identify with. Our own beliefs tend to be converted into *truth* much more easily than is warranted. Even in the most abstract of questions, it's easy to see, in oneself or in others, a strength of belief that the world *is* a certain way—material or ideal, free or fated, inherently good or evil—all out of proportion to the actual evidence and even when confronted with many people believing quite different things for reasons they find just as persuasive. This is a tendency worth keeping a keen eye on.

But it isn't *just* tribalism. Convincing, opposing views persist because the questions are *hard*. The easy ones get settled and largely forgotten about. In a well-run government, only the *hard* questions come to the desk of the emperor (or president). So too with the structure we've built out of our hard-won understanding of the world, both individually and collectively. Of course, sorting things into what has been settled and what is still unknown or open for debate is not always easy, or agreed upon, or widely disseminated throughout society. But overall, there is some sense of a cutting edge: the world really is not flat, but we can debate about the speed at which it is warming; lightning is caused by segregation of electric charges, not gods' whims; gravity is not a *propensity to go down* but rather a *propensity for stuff to curve space-time in a particular way*. Knowing exactly where this cutting edge lies can be something of an art. Sometimes it is obvious and surrounded by debate. Sometimes, as we've seen, it can be much more subtle, discoverable only through a careful skepticism of simple or familiar answers. But there it lies, still licking its chops.

Even knowledge of *moral* truths, which are perhaps harder to place on the subjective-objective axis of being, has progressed. Slavery is wrong; discrimination is wrong; women are equal in worth to men; might does not make right. Even if they contain some invention as well, we must choose to see these as hard-earned discoveries that we cannot unknow. We may then focus on how to bring our reality in closer accord with these ideals, as well as tackle the still much more open questions: What makes a just society? How do we balance liberty and security? When does technology bring, or undermine, well-being? What does the future hold, with its quantum computers, artificial intelligences, genetic engineering, and interplanetary travel—and how can we make that a future we really want?

The progress we've made so far comes from a long history of remarkable human effort and drive. Humans are agents of action. Minute to minute, hour to hour, week to week, we decide, decide, decide, and act, act, act. Our cognitive architecture is that of an astonishing prediction-and-decision engine, much of which runs with little notice. We dash across rocks, make coffee, and shuffle papers more or less on autopilot. But when the predictions or the decisions get *hard*, then suddenly they are at the center of our awareness. This can become very stressful, frightening even. But it is also often when we feel most alive, most interested, most awake. Who wants to read a novel or see a drama in which no character has to make a difficult decision, caught in a dilemma between conflicting desires and with only guesses as to what will come?

Yet humans don't just act and decide and feel. We have also learned to contemplate. To think and ponder free-style on ideas and truths that may have elegance and beauty but near-zero utility. I'm often amazed that we can do this at all, let alone so well as to conjure an understanding of the metric structure of space-time, or the evolution of the quantum state, or the meaning and generation of entropy. And we've learned to contemplate in other less intellectual ways. To very simply and purely *be*. To look ever more closely at our own minds and how they run. To see the minute gradations of our relations with others and with the world.

Like adventures in the physical world, this contemplation of the inner world and the world writ large can be exhilarating and even terrifying. As one's mental picture of the Universe and its operation shifts, reality can swing from being a comforting den to a windswept mountaintop to a dark forest and back again. But you're made of that very, very special stuff, infinitesimal in the universe, that is able to appreciate this, and to take the adventure. The journey can be exhausting, true, painful, humbling, but I often feel: "What could be better?"

Who am I?

What is this?

Where, from here?

51

THE ARROW

The arrow approaches, along all possible trajectories.
The arrow is still, as your mind moves across space and time.
You watch as snow falls gracefully on the Lhasa River.
A tear glistens on the sensei's cheek.
A distant temple bell rings.
Lotus flowers endlessly open.
Can it be avoided? Even if the fight is in vain, it is worthwhile.
        Better even.
In the expanse of a heartbeat, ineffable reality is asked and
        answers an infinity of questions.
Layers of mist conceal and cool the desert.
Turned around by the eyes of samadhi, a die is uncast.
A dove takes flight, and falls.
All paintings dissolve, and are redrawn.
A net untangles.
In the blue sky—boundless, endless, hidden—the stars beckon.
You step through the gates.

ACKNOWLEDGMENTS

It's been a long road that started (but of course didn't) with a walk in the snow and ended (but of course didn't) with the completion of this book, and I'm incredibly grateful for the company I've had. Michael Batshaw provided the seed idea for a book of koans, along with encouragement, insight, and support along the way. Sally Hurley (then Aguirre), for just as many years gave endless inspiration, patience, and love. And a good dose of Zen.

As the book started to come together, I was fortunate to have Ben Miller, Amita Kuttner, Sally, and Brad Hurley heroically work through early versions of the book and give detailed feedback. I'm also grateful to my parents, as well as Peter Gregorio, Kokyo Henkel, Jenann Ismael, Max Tegmark, and Jan Wallaczek, for reading, feedback, and much-needed moral support.

Max and John Brockman, as well as Matt Weiland, Jeff Shreve, Remy Cawley, and others at Norton, took a chance on this rather unusual idea for a book, and I'm grateful for their support and suggestions.

It's been a true pleasure to work with the talented Zach Corse, who patiently and insightfully created an amazing set of figures and illustrations. Copyediting with Stephanie Hiebert was a pleasure and an education.

I would also like to acknowledge the local part of the unfettered mind that it's been an honor, adventure, and privilege to be a part of. Thanks to David Layzer, for in many ways setting me on the path of a lot of the questions in this book. To Max Tegmark, for years of fascinating discussions, successful schemes, and everything in-between. To the Templeton Foundation and the Foundational Questions Institute (and so many people I've gotten to know through both!), and UC Santa Cruz, for supporting my research and thinking even when somewhat audacious. Behind them, my gratitude to all those who struggled and supported those who struggled to ask the questions. Ahead of us all, I bow to the unknown future to be made.

NOTES

Introduction

1. As quoted in Ludwig von Bertalanffy, *Problems of Life: An Evaluation of Modern Biological Thought* (Eastford, CT: Martino Fine Books, 2014), 1.
2. Michael White and John Gribbins, *Einstein: A Life in Science* (London: Simon & Schuster, 1993), 262.

Part 1: The Path Laid Out before Us

1. As translated by Jay Garfield in *The Fundamental Wisdom of the Middle Way: Nāgārjuna's Mūlamadhyamakākrikā* (New York: Oxford University Press, 1995), 6. Nāgārjuna was a second-century Buddhist philosopher who lived in India, and whose thought and writing forms the foundation of Mahāyāna Buddhism, of which Zen Buddhism is one subset.
2. William James, *The Principles of Psychology* (New York: Dover, 1918), 608.
3. Galileo Galilei, *Dialogue concerning the Two Chief World Systems, Ptolemaic & Copernican*, trans. Stillman Drake (Berkeley: University of California Press, 1953), 187.
4. To be more precise, the Sun is moving at 370 km/s (accurate to 1%) relative to the *cosmic microwave background* radiation pervading space as measured by satellite experiments; see C. H. Lineweaver et al., "The Dipole Observed in the COBE DMR 4 Year Data," *Astrophysical Journal* 470 (1996), 38.
5. As quoted in Stillman Drake, *Galileo at Work: His Scientific Biography* (Mineola, NY: Dover, 1978), 186.
6. "Einstein Is Found Hiding on Birthday: Busy with Gift Microscope," *New York Times*, March 15, 1929, 3.

7. Kazuaki Tanahashi, ed., *Moon in a Dewdrop: Writings of Zen Master Dōgen*, trans. Robert Aitken et al. (San Francisco: North Point Press, 1985), 77.

8. Rolling objects are actually a bit more complex to treat correctly than is depicted here, but for present purposes, their behavior is essentially the same as if they were sliding. Galileo actually used rolling balls also in his experiments, eliding some of this complexity as well. As a special bonus for those with some training in Newtonian mechanics and a propensity to read endnotes, here is a fun paradox: Why does a rolling ball ever stop rolling? For the center of mass to slow, there must be a force (from the ground) acting against the direction of the ball's overall motion. Yet this force is applied to the edge of the ball, and hence constitutes a *torque*, in a direction that should cause the ball to rotate faster. But faster rotation means faster rolling. How does this make sense?

9. Since Galileo's time, scientists have verified the equivalence of inertia and gravitational charge to a precision of better than one part in a trillion; see C. M. Will, "The Confrontation between General Relativity and Experiment," *Living Reviews in Relativity* 4, no. 1 (2001): art. 4.

10. Measuring accurate distances and directions is far more difficult than our time of GPS and odometers would suggest. But since that's not the most interesting problem, let's suppose that the khan's talented riders and scholars were able to overcome this challenge and make their measurements with very high accuracy.

11. This tool—a *scale* that varies from place to place and takes some map distances (or *coordinate* distances) and converts them into a real physical distance—is known in more technical parlance as a *metric*. The only real difference is that, in general, the metric can also incorporate *products* of the north–south map distance and the east–west map distance.

12. Isaac Newton, *Newton's Principia: The Mathematical Principles of Natural Philosophy*, trans. Andrew Motte (New York: Daniel Adee, 1846), 77.

13. As quoted by Eihei Dōgen in Tanahashi, *Moon in a Dewdrop*, 131. Samadhi is a meditative state of extreme concentration.

14. If the gondola and ball are moving in different directions, the exact same rule holds, except that we must consider vector velocities (which contain both the speed and the direction of the objects), and these velocities must be mathematically added as vectors.

15. A nice review of the role of both special and general relativity in the GPS is Neil Ashby's "Relativity and the Global Positioning System," *Physics Today* 55, no. 5 (2002): 41–47.

16. Quoted in John Gribbin, *Einstein's Masterwork: 1915 and the General Theory of Relativity* (New York: Pegasus Books, 2016).

17. This terminology is commonly used because these forces don't appear to be caused by any sort of interaction with other stuff (as in friction, contact, or magnetic forces) or by any fundamental force such as electromagnetism.

18. In fact, this mathematics is so elegant that it can describe space of 4, 5, 11, or any other number of dimensions with equal ease! It can also describe even more general mathematical spaces with no known relation to that which we inhabit.

19. Abraham Pais, *"Subtle Is the Lord—": The Science and the Life of Albert Einstein* (Oxford: Clarendon, 1982), 216.

20. Max Born, *Physics in My Generation* (New York: Springer, 1968), 109.

Part 2: An Uncertain Trail through Treacherous Terrain

1. Modern circuits also incorporate devices that make use of quantum mechanical effects—but the behavior of these devices is also quite predictable, so the essential point is unchanged.

2. As translated and quoted by David Layzer in *Cosmogenesis: The Growth of Order in the Universe* (New York: Oxford University Press, 1990).

3. Quoted in Walter Isaacson, ed., *Einstein: His Life and Universe* (New York: Simon & Schuster, 2008), 540.

4. As for the Greek ideas, however, this does not appear to be an absolutist doctrine, since there also appears to be a notion of modes of being that are outside of causes and conditions.

5. Nick Herbert, *Quantum Reality: Beyond the New Physics* (New York: Anchor Press/Doubleday, 1987), 53.

6. If you're troubled that after this procedure the path next to Adyar has just 1/16 probability, you're right to be! We have to be very careful what we assume. Here, the 75%-25% probability split was placed at the fork north of Qinghai (then the probability was evenly split at each fork going west). Had we assumed that Adyar's path had a 25% probability, we would have been forced into a different set of probabilities, attributing, for example, 0% probability to the path just south of the Taklimakan desert. This difference points to a quite subtle general problem of how to assign a "measure," or set of probabilities, to a set of choices when there is not a single obvious way.

7. Gottfried Wilhelm von Leibniz, *The Monadology: And Other Philosophical Writings*, trans. Robert Latta (n.p.: Oxford University Press, 1898), 62.

8. An interesting version of a quantum die, which I've used here, would be a spin 5/2 particle. Such a particle would give only the answers +5/2, +3/2, +1/2, −1/2, −3/2, −5/2 to the question "how fast are you spinning about a vertical axis?" (or any other axis). These spins might be mapped onto (*6, 5, 4, 3, 2, 1*) facing in the direction of that axis to provide a die analog. (My thanks to Steven Gratton for fun discussion of this model—and some numbers.)

9. In a bit more detail, each of these coefficients is a complex number (complex numbers will be discussed later) and must be squared to obtain the probability.

10. Technically, this approach sneaks in an assumption: that if we measure a prop-

erty of a system and then immediately measure the same property again, we will get the same answer.

11. To multiply two such complex numbers, you multiply their magnitudes and add their phases, keeping in mind that if the phase exceeds 360 degrees, it wraps around to 0 degrees. A second way to think of a complex number is as two numbers (a *real number* and an *imaginary number*) bundled together— something like coordinates such as longitude and latitude.

12. This effect underlies Heisenberg's famous *uncertainty principle*: that you cannot precisely know both the position and the velocity of a particle at the same time. One can look at this as a particular position (state) being composed of many different velocity states, and vice versa.

13. It is a crucial property of quantum mechanics that if you have two solutions to Schrödinger's equations, then when you add them together you get a new solution. This property allows individual treatment of separate systems, which can then be combined into a single system (if they coincide in space-time) by simple addition.

14. E. Joos and H. D. Zeh, "The Emergence of Classical Properties through Inter-action with the Environment," *Zeitschrift für Physik. B, Condensed Matter* 59, no. 2 (1985): 223–43.

15. Describing them all would take a book itself. Worth a mention are the *dynam-ical collapse* view and the *hidden variables* view; both are rather different from either the epistemic or ontic described, but they are significantly less popular.

16. Jean-Paul Sartre, *The Flies and In Camera*, trans. Stuart Gilbert (London: H. Hamilton, 1946), 74.

17. A. M. Turing, "On Computable Numbers, with an Application to the Entscheidungsproblem," *Proceedings of the London Mathematical Society* s2-42, no. 1 (1937): 230–65.

18. David J. Chalmers, *The Conscious Mind: In Search of a Fundamental Theory* (New York: Oxford University Press, 1996), 1.

19. In an amusing example, when I reread this paragraph I found that I had typed "with a significant stochastic component entertaining our thoughts." I have no memory of writing that!

20. Turing, "On Computable Numbers." Note that *if* your mind is equivalent to a Turing machine, this means that *you* cannot, in general, know whether a program will halt. This could be page four of *What Cannot Be Known*, but I'll let you write the poem.

21. About 12.8 billion seconds later, computer scientist Seth Lloyd would put together the same set of implications; see "A Turing Test for Free Will," October 11, 2013, https://arxiv.org/pdf/1310.3225.pdf.

Part 3: Torn Apart and Reassembled

1. Technically, the logarithm of this count.

2. This is rather subtle, because as we have seen, we can create a superposition of any two quantum states, and a continuously valued complex number can multiply each element of the superposition; this means that there are an infinite number of seemingly different states we could write down. Nonetheless, there is generally a finite amount of information to be extracted from any such state, and it is more useful to classify a quantum system in terms of the number of distinguishable outcomes that a complete measurement could find.

3. A useful way to describe this is in terms of total energy and free energy. *Total energy*, which is conserved, includes *all* forms of energy in a physical system. *Free energy*, on the other hand, is the amount of the total energy that is actually available to do useful work. For example, a box of hot gas in equilibrium, by itself, cannot do work: the system has already attained maximal disorder, so its natural evolution to anything else would be stupendously improbable. However, if a hot box of gas is coupled to a cold box of gas (with equal gas density), then work can be done: the hot gas pushes into the cold gas and could, for example, drive a turbine. In the first case there is plenty of energy (the gas is hot), but little or no *free* energy. In the second case there is a bit more energy (we've added the energy of the cold gas), but also lots of free energy. The energy that comes from the Sun is "useful" energy: it contains lots of free energy that we can do things with—and ultimately this is what keeps the Earth so far away from equilibrium chaos.

4. Specifically, we could write this as $h(t) = h(t_0) - (\frac{1}{2})g(t - t_0)^2$, where g is 9.8 m/s per second and h is the height.

5. "The Landscape: A Conversation with Leonard Susskind," *Edge*, December 2, 2003, https://www.edge.org/conversation/leonard_susskind-the-landscape.

6. There is also a quantum version, called *von Neumann entropy*, that behaves quite similarly.

7. There are, of course, more. Six general classes are listed in the *Stanford Encyclopedia of Philosophy*, s.v. "Intepretations of Probability," accessed 2/11/18, https://plato.stanford.edu/entries/probability-interpret. The *Bayesian* and *frequentist* views are the most widespread, however.

8. As it turns out, there is a mathematical object into which these two notions of uncertainty can both be packaged. Called a *density matrix*, it is the tool of choice for quantum physicists and djinns alike.

9. Technically, Shannon's definition takes a set of probabilities P_i, where $i = 1 \ldots N$, and defines a randomness R as the sum over i of $-P_i \log_2 P_i$.

10. There is a slight generalization, devised by von Neumann, that applies in quantum mechanics.

11. Specifically, this surprisingly novel quantity is the sum over i of $-P_i \log P_i/N_i$

with P_i and N_i being respectively the probability of, and number of micro-states in, the ith macrostate.

12. The definitions of R and D are related in that if we choose each macrostate to be simply a microstate, the definitions are the same.

13. The idea that gathering information *must* generate entropy goes back to Maxwell's time, where he envisioned a creature that could measure the speed of individual molecules and, by doing so, sort a warm box of gas into hot and cold halves, thus decreasing its entropy. The idea has been used since then to derive many connections between information theory and thermodynamics.

14. The math with these exponents gets tricky. The key point is that $(X^Y)^Z = X^{Y \times Z}$, which is different from $X^{(Y^Z)}$. So here, $(10^{10})^{10^{26}} = 10^{(10 \times 10^{26})} = 10^{10^{27}}$.

15. In the first stage, the warm gas in the box would eventually break down the painting's material; in the second, the chemical energy built into the organic compounds—especially paper—that the painting is made of would be released, effectively burning the painting in slow motion. There would then be a phase of ultraslow nuclear fusion to make the gas even hotter. And eventually, the protons in the gas would probably decay into quarks, leaving a soup of photons, quarks, electrons, and neutrinos.

16. This result was first devised by the French mathematician Henri Poincaré in the context of classical mechanics, but it applies even more broadly.

17. That is, anything in the state space of the system, with the same conserved quantities. This statement is not quite true at the quantum level, in that you can show that for a given initial state, there are other states (compatible with conserved quantities) that you can get only so "close" to.

18. For a technical but somewhat accessible review, see *Stanford Encyclopedia of Philosophy*, s.v. "The Consistent Histories Approach to Quantum Mechanics," accessed 11/27/17, https://plato.stanford.edu/entries/qm-consistent-histories.

19. *Stanford Encyclopedia of Philosophy*, s.v. "The Consistent Histories Approach to Quantum Mechanics."

Part 4: Lofty Peaks with Endless Views

1. Giordano Bruno, *De l'infinito universo et mondi* [On the infinite universe and worlds], in Dorothea Waley Singer, *Giordano Bruno, His Life and Thought* (New York: Schuman, 1950).

2. See Edward Harrison, *Darkness at Night: A Riddle of the Universe* (Cambridge, MA: Harvard University Press, 1987) for more, including a fairly comprehensive table of proposed solutions. Most of these fail, but there is a rather tricky, subtle, and interesting way, going back to John Herschel, to have a model universe that is infinitely old, and large, and static, and filled with infinitely old, blazing-away stars. It's even possible for them to be arranged in a way that

could be called uniform, but not in the usual sense. If you're interested, consult Benoit Mandelbrot's 1982 book *The Fractal Geometry of Nature*.

3. The infinite set of stars does generate a radiation bath, but at a finite temperature that is set by a competition between the brightness and creation rate of the stars, and the expansion of space that dilutes the radiation away. And while it's true that there is a star in every direction, the light from very distant stars is redshifted and cooled enough by the cosmic expansion to add up to a finite total amount of radiation seen by any given observer.

4. Imagine, for example, stretching an 8×11 sheet of rubberized graph paper, with 10 lines per inch, into a 16×22 piece, during 1 second. Two neighboring lines would separate by $\frac{1}{10}$ inch per second, whereas lines 1 inch—10 times as far—apart would separate by 1 inch per second, or 10 times as fast.

5. In uncurved space-time it is literally a sphere. More generally, it has spherical topology.

6. Guth was not the only one thinking along these lines, or even the first to publish about exponential expansion (that was Alexei Starobinsky)—but it's fair to say that he synthesized the argument for cosmological inflation first, and so elegantly that it is often presented in almost exactly the same way even now.

7. See, for example, Max Tegmark et al., "Dimensionless Constants, Cosmology and Other Dark Matters," *Physical Review. D* 73 (2006): 023505.

8. B. J. Carr and M. J. Rees, "The Anthropic Principle and the Structure of the Physical World," *Nature* 278 (1979): 605; see also Victor F. Weisskopf, "Of Atoms, Mountains, and Stars: A Study in Qualitative Physics," *Science* 187, no. 4177 (1975): 605. For a popular and readable account, see Martin Rees's *Just Six Numbers: The Deep Forces That Shape the Universe* (New York: Basic Books, 2000).

9. If we think more speculatively about very different forms of life (living, say, on the crust of a neutron star, or built out of black holes), then things are quite a bit less restrictive.

10. Gottfried Wilhelm von Leibniz, *The Monadology: And Other Philosophical Writings*, trans. Robert Latta (n.p.: Oxford University Press, 1898), 247.

11. It's notable that it takes 18 weeks to go from one plant to a large but tractable area, but in another 12 weeks the plant growth would be completely and hopelessly out of control. This is generic behavior of exponential growth: it can run under the radar for a while, then suddenly explode into view. It's a good thing to keep in mind whenever you hear that something (population, computing power, etc.) is growing exponentially.

12. The fact that the observed vacuum energy is very tiny (about 10^{-122} of the maximum possible energy density) but nonzero is known as the *cosmological constant* problem and is one of the biggest mysteries in cosmology.

13. The pattern is very interesting, actually being *fractal*. In particular, although the inflating volume grows exponentially, as time goes on it is a smaller and smaller fraction of what it would have been without the inflation-ending process. Thus, in the long time limit, formally a *zero* fraction of space is still inflating, even while it has immense volume!

14. One can alternatively see general relativity as being an implication of string theory, but in any event, the basic structure of general relativity is taken to hold across the space-time created by eternal inflation. Quantum theory, on the other hand, is generally taken to be truly universally valid. Why? That's a fair question.

15. Parts of this koan are taken almost verbatim from Galileo's original dialogue (see note 1-3), it seeming both hopeless and pointless to try to improve upon them!

16. However, as was demonstrated 264 years ago by Georg Cantor (who was literally driven mad, perhaps by infinity), some infinite sets are bigger than others. In particular, the set of all possible *sets of integers* is mathematically demonstrably bigger than the set of all integers; one cannot create a two-way mapping between the two sets.

17. Don't take this too seriously. But it's true that a single point can, *in principle*, be in causal contact with an infinite uniform space in this construct.

18. For a more detailed argument, see Anthony Aguirre and Max Tegmark, "Born in an Infinite Universe: A Cosmological Interpretation of Quantum Mechanics," *Physical Review. D* 84 (2011): 105002.

19. Kamala Masters, "The Preciousness of Our Human Life," accessed 4/8/18, http://vipassanametta.org/wp/wp-content/uploads/2012/01/The-Preciousness-of-Our-Human-Life.pdf.

20. For an extensive treatment of this set of ideas and their applications to cosmological and other questions, see Nick Bostrom, *Anthropic Bias: Observation Selection Effects in Science and Philosophy* (New York: Routledge, 2002). It's full of delightful (i.e., interesting and sometimes troubling!) thought experiments; check out, for example, the "doomsday" and "Sleeping Beauty" paradoxes.

21. We also have lots of room to grow, with enough heavy elements in our solar system for about 10^{26} people (using the elements at an unrecommended maximum efficiency), and an observable universe with about 10^{24} times as many resources as that. More "people" might, in principle, be packed in if those people were posthuman entities instantiated in some other computational medium, but let's not go there.

Part 5: Who Am I? Don't Know!

1. As of 2017, estimates are on the order of 10^{10} terabytes in the "datasphere."

2. Technically, you can always choose your "basis" of states so that the system is in just one of them. But as we saw in WHAT CANNOT BE KNOWN, there's no way to know what that basis is!

3. You can do this right now: online sources like RandomNumbers.info (http://www.randomnumbers.info) provide genuinely quantum outcomes, and you can use them to place bets, if you like.

4. Even worse, 1 versus $\sqrt{2}$ would require infinitely many of each outcome to get right! This view gets even more troubling if you try to square it with any moral philosophy in which it is good to cause the existence of many happy people, and in which people have equal intrinsic moral worth; see, for example, Rachael Briggs and Daniel Nolan, "Utility Monsters for the Fission Age," *Pacific Philosophical Quarterly* 96, no. 3 (2015): 392–407.

5. There is no consensus among adherents of the ontic view on how to think about the probabilities. I've always been deeply skeptical of the many-worlds account for this reason. For the closest I've come to making sense of the probabilities via a rather strange route, see Anthony Aguirre and Max Tegmark, "Born in an Infinite Universe: A Cosmological Interpretation of Quantum Mechanics," *Physical Review. D* 84 (2011): 105002.

6. Probably GUT-scale or Planck-scale—many, many, many orders of magnitude higher than anything technologically feasible today, in case you're contemplating this.

7. The probability of this happening in, say, a cubic meter during a second is determined by the physics governing the decay process, just as in a radioactive atom. *If* such a decay has not occurred during the past 13.8 billion years, then it is unlikely to happen tomorrow.

8. This branching is intimately related to the second law of thermodynamics. Decoherence evolves a pure state into something that we then choose for convenience to describe as a probabilistic mixture of two different states, but that corresponds to higher disorder and randomness. "Recoherence" would correspond to a reduction in randomness or disorder, and can be engineered but will essentially never happen on its own.

9. Andy Clark and David J. Chalmers, "The Extended Mind," *Analysis* 58, no. 1 (1998): 7–19.

10. It's easy to forget how very long prehistory was. A hundred thousand years is 5,000 generations. Even with a global population of just millions, that means there are billions of humans who lived for which we have essentially no direct record.

11. Other than the last line, Sōhō's and Munenori's statements are directly quoted from their books: Takuan Sōhō, *Unfettered Mind: Writings of the Zen Master*

to the Sword Master, trans. William Scott Wilson (Tokyo: Kodansha International, 1986), 19, 26; Yagyū Munenori, *The Swordsman's Handbook: Samurai Teachings on the Path of the Sword*, trans. William Scott Wilson (Boston: Shambhala, 2012), 25.

Part 6: Form Is Emptiness; Emptiness Is Form

1. One of the most important uses of quantum computers is that they can efficiently simulate physical quantum systems. We can accomplish such simulations by setting up a sort of mapping between elements of the quantum computer and of the quantum system in question, then evolving the quantum computer to understand what happens to the quantum system. There's a real sense in which simulation and reality are the same thing.

2. Steven Weinberg, "What Is an Elementary Particle?" *Beam Line*, 1996, http://www.sbfisica.org.br/~evjaspc/xvi/arquivos_diversos/27-1-weinberg.pdf.

3. Leonhard Euler, "A Demonstration of a Theorem on the Order Observed in the Sum of Divisors," July 30, 2009, https://arxiv.org/pdf/math/0507201.pdf.

4. Wikipedia, s.v. "Johannes Zukertort," accessed 12/6/17, https://en.wikipedia.org/wiki/Johannes_Zukertort.

5. Fred Child, "Unknown Bach Aria Discovered in Germany," NPR, June 9, 2005, http://www.npr.org/templates/story/story.php?storyId=4695336.

6. Karl Popper, "Three Worlds" (Tanner Lecture on Human Values, University of Michigan, April 7, 1978).

7. Energy conservation is fundamentally a result of the laws of physics being independent of time. But in an expanding universe, the laws governing the motion of particles are *not* time independent, because the space in which they exist is changing with time.

8. This is a long and very interesting story that goes untold here. For an in-depth and entertaining discussion, you might try Lenny Susskind's *The Black Hole War: My Battle with Stephen Hawking to Make the World Safe for Quantum Mechanics* (New York: Little, Brown, 2008).

9. How does this square with energy and other types of conservation? In a typical quantum way. In quantum theory, energy (for example) is a quantity, or property, that you can observe. A superposition of macroscopic objects would correspond to two different objects with two different energies, but if you measure the energy, you won't get the *sum* of those energies (as if you actually physically copied the object), but rather one or the other.

10. See Metamath Proof Explorer, "Theorem 2p2e4," accessed 12/2/17, http://us.metamath.org/mpegif/2p2e4.html.

11. More precisely, this *maximum entropy* technique would assign probabilities to states based on the set of probabilities that maximizes the entropy (by some definition) computed from those probabilities.

12. This is not a bomb to be trifled with. Replacing ordinary matter with a gas of elementary particles would liberate much more heat energy than nuclear fusion does, so such a bomb would be far more powerful than any nuclear weapon ever devised.

13. Computing elements are more competitive but still much bulkier than functional biomolecules. As of 2018, about 50 million transistors can be packed into a square millimeter, but you could fit about a million bacteria into that same area.

14. J. A. Wheeler, "Frontiers of Time," in *Problems in the Foundations of Physics* (Proceedings of the International School of Physics "Enrico Fermi," Course 72), ed. G. Toraldo di Francia (Amsterdam: North-Holland, 1979), 395–492.

15. For an excellent and fairly readable account, see Samuel Rathmanner and Marcus Hutter, "A Philosophical Treatise of Universal Induction," *Entropy* 13, no. 6 (2011): 1076–1136.

16. Technically, weighting by 2^{-L}, where L is the length of the program.

17. For more on this, see the papers by Markus Müller, who applies (generalizations of) Solomonoff's induction to create a quite fleshed-out version of Wheeler's "law without law"—for example, Markus P. Müller, "Could the Physical World Be Emergent Instead of Fundamental, and Why Should We Ask? (Short Version)," December 5, 2017, https://arxiv.org/pdf/1712.01816.pdf.

18. The actual historical figures and references to their quoted words are, in order: **Buddha**: Dwight Goddard, ed., *A Buddhist Bible* (Boston: Beacon Press, 1994), 89; **Berkeley**: George Berkeley, "The First Dialogue," in *Three Dialogues between Hylas and Philonous in Opposition to Sceptics and Atheists*, Harvard Classics, no. 37, pt. 2 (New York: Collier, 1909–14; New York: Bartleby.com, 2001), https://www.bartleby.com/37/2/1.html; **Nāgārjuna**: Jay Garfield, trans., *The Fundamental Wisdom of the Middle Way: Nāgārjuna's Mūlamadhyamakākārikā* (New York: Oxford University Press, 1995), 49; **Descartes**: Wikisource, s.v. "Meditations on First Philosophy/Meditation I," accessed 7/28/18, https://en.wikisource.org/wiki/Meditations_on_First _Philosophy/Meditation_I; **Schopenhauer**: Arthur Schopenhauer, *The World as Will and Idea*, trans. R. B. Haldane and J. Kemp, 6th ed. (London: Kegan Paul, Trench, Trübner, 1909), 2:166, Project Gutenberg, http://www.gutenberg.org/ebooks/40097?msg=welcome_stranger; **Eddington**: Arthur Stanley Eddington, *The Nature of the Physical World* (Cambridge: Cambridge University Press, 1929), 277–78, Project Gutenberg Canada ebook no. 1097, https://gutenberg.ca/ebooks/eddingtona-physicalworld/eddingtona-physicalworld-01-h-dir/eddingtona-physicalworld-01-h.html.

19. Eddington, *Nature of the Physical World*, 281, 280.